Placer Gold Deposits of Arizona

By MAUREEN G. JOHNSON

GEOLOGICAL SURVEY BULLETIN 1355

A catalog of location, geology, and production with lists of annotated references pertaining to the placer districts

UNITED STATES GOVERNMENT PRINTING OFFICE, WASHINGTON : 1972

UNITED STATES DEPARTMENT OF THE INTERIOR

ROGERS C. B. MORTON, *Secretary*

GEOLOGICAL SURVEY

V. E. McKelvey, *Director*

Library of Congress catalog-card No. 72-600097

For sale by the Superintendent of Documents, U.S. Government Printing Office
Washington, D.C. 20402 - Price $1 (paper cover)
Stock Number 2401-2155

CONTENTS

ILLUSTRATIONS

TABLES

PLACER GOLD DEPOSITS OF ARIZONA

By Maureen G. Johnson

ABSTRACT

Eighty-seven placer districts in Arizona are estimated to have produced a minimum of 564,052 ounces of placer gold from 1774 to 1968. The location, areal extent, past production, mining history, and probable lode source summarized for each district are based on information obtained from a wide variety of published reports relating to placer deposits. Annotated references to all reports that contain information about individual deposits are given for each district.

Most of the placer gold found in Arizona was derived from systems of small gold-quartz veinlets and stringers scattered throughout the bedrock of the adjacent mountain ranges. In only a few localities was the gold in large placer deposits derived from well-defined vein systems mined for the lode-gold content. The most productive placer mining era was 1858–80, when rich deposits of placer gold were found in the southwestern and central parts of Arizona and hundreds of individuals worked the rich surface accumulations of gold. Subsequent placer mining was done by individuals working small deposits in many areas of the State using small-scale portable equipment, such as rockers, sluices, and drywashers. Large-scale dredge operations were active in a few districts.

INTRODUCTION

HISTORY OF PLACER MINING IN ARIZONA

Arizona's placer-mining industry began in 1774, when Padre Manuel Lopez reportedly directed Papago Indians in mining the gold-bearing gravels along the flanks of the Quijotoa Mountains, Pima County. Placer mining was active in that region from 1774 to 1849, when the discovery of gold in California apparently attracted many of the Mexican miners who worked the gravels (Stephens, 1884). Arizona was then part of Mexico, and little is known of the placer mining that probably was carried on in various parts of southern Arizona. Placers were probably worked in the Oro Blanco district, Santa Cruz County, and the Arivaca district, Pima County. The part of Arizona north of the Gila River was ceded to the United States in 1848, and the part of Arizona south of the Gila River, where most of the early placer mining occurred, was purchased in 1853. Placers were discovered in the 1850's in the Bagdad area, Yavapai County, and Chemuehuevis Mountains, Mohave County; but it was not until 1858, when placers were discovered by Colonel Jacob Snively at the north

1

end of the Gila Mountains, Yuma County, that the first placer-mining rush in Arizona was precipitated. The early years of the 1860's saw the discovery of the famous placers at La Paz, Yuma County, and Rich Hill and Lynx Creek, Yavapai County; many smaller and less famous placer fields were discovered at that time.

In the 1860's, Arizona was a relatively isolated and underpopulated territory, fraught with communication and travel difficulties, and beset by Indian problems. Placer mining was actively pursued throughout the territory, and some rich lode-gold mines were discovered and worked; but real news of Arizona mining was slow to filter out from the territory to the more populated areas in California and the East. The period from 1860 to 1880 is reported as the most active and productive period in placer mining, but because of poor communications, there is very little reliable information or production record.

By 1900 most placer areas had been discovered, and many were nearly worked out. Placer mining continued intermittently during the early years of the 1900's. Many attempts were made in various parts of the State to mine placer gravels by drywashing machines, but it was not until the economic impetus of the depression that placer mining became active again in Arizona. During the years 1930–38, 95 different districts were credited with placer gold production, but many of these districts produced only a few ounces.

After the boom of the 1930's, the war years of the 1940's were a setback to gold mining activity. War Production Board Order L–208 greatly restricted the development of gold mines; prospecting for and mining metals essential to the war effort was deemed more important than mining gold. Even more important, however, the economy of the 1940's encouraged work in offices, factories, and war industries for those not in military service, and as a result, many miners and prospectors left the gold fields and never returned.

After 1942, placer production never again reached the heights of the 1930's or the peak production of the 1860's to 1880's.

PURPOSE AND SCOPE OF PRESENT STUDY

The present paper is a compilation of published information relating to the placer gold deposits of Arizona, one of a series of four papers describing the gold placer deposits of the Southwestern States. The purpose of the paper is to outline areas of placer deposits in the State of Arizona and to serve as a guide to their location, extent, production history, and source. The work was undertaken as part of the investigation of the distribution of known gold occurrences in the Western United States.

Each placer is described briefly. Location is given by geographic area and township and range. (See pl. 1 for location of the placer gold deposits.) Topographic maps and geologic maps that show the placer area are listed. Access is indicated by direction and distance along major roads and highways from a nearby center of population.

Detailed information relating to the exact location of placer deposits, their thickness, distribution, and average gold content (all values cited in the text have been converted to gold at $35 per ounce, except where otherwise noted) is included, where available, under "Extent." U.S. Bureau of Land Management land plats of surveyed township and ranges were especially helpful in locating old placer claims and some creeks and drywashes not named on recent topographic maps. These land plats were consulted for all the surveyed areas in Arizona and were especially useful in locating placers in Yavapai County, for which recent topographic maps are not available, although most of the area is surveyed. U.S. Bureau of Mines records were also consulted for the location of small placer claims.

Discovery of the placer deposit and subsequent placer-mining activity are briefly described under "Production history." Detailed discussion of mining operations is omitted, as this information can be found in the individual papers published by the State of Arizona, in the yearly Mineral Resources and the Mineral Yearbook volumes published by the U.S. Bureau of Mines and the U.S. Geological Survey, and in many mining journals. Placer gold production, in ounces (table 1), was compiled from the yearly Mineral Resources and Mineral Yearbook volumes and from information supplied by the U.S. Bureau of Mines. These totals of recorded production are probably lower than actual gold production, for substantial amounts of coarse placer gold commonly sold by individuals to jewelers and specimen buyers are not reported to the U.S. Bureau of Mines or to the U.S. Bureau of the Mint. Information about the age and type of lode deposit that was the source of the placer gold is discussed for each district.

An extensive body of literature was searched to find information relating to all the placers in the State. A list of literature references is given with each district with annotation indicating the type of information found. Sources of information are detailed reports on mining districts, general geologic reports, Federal and State publications, and brief articles and news notes in mining journals. The Arizona Bureau of Mines has published a series of bulletins describing the geographic location and history of many placers in the State (Wilson, 1961, and earlier editions). The present report, which draws on much information contained in the State bulletins, emphasizes locations keyed to topographic maps, detailed production data, and an extensive bibliography. A complete bibliography, given at the end of the paper, includes separate sections for all literature references and all geologic map references.

Map publications of the Geological Survey can be ordered from the U.S. Geological Survey, Distribution Section, Denver Federal Center, Denver, Colo. 80225; book publications, from the Superintendent of Documents, Government Printing Office, Washington, D.C. 20402.

COCHISE COUNTY

1. DOS CABEZAS AND TEVISTON DISTRICTS

Location: North and south flanks of the Dos Cabezas Mountains between San Simon Valley and Sulphur Spring Valley, Tps. 13–15 S., Rs. 26 and 27 E.

Topographic maps: Dos Cabezas and Luzena 15-minute quandrangles.

Geologic map: Cooper, 1960, Reconnaissance map of the Wilcox, Fisher Hills [now named Luzena], Cochise, and Dos Cabezas quadrangle, scale 1:62,500.

Access: The Dos Cabezas district, on the south flank of the Dos Cabezas Mountains, is 15 miles east of Willcox on State Highway 186. The Teviston district, on the north side of the Dos Cabezas Mountains, is accessible by a road leading 5 miles south from Interstate 10, 17 miles northeast of Willcox and 6 ½ miles west of Bowie (formerly Teviston).

Extent: Placers on the south flank of the mountains (Dos Cabezas district) are said to be found in all the gulches draining the mineralized part of the mountain range, an area about 3 miles long between Walnut Canyon and Howard Canyon. The most actively worked placers are located in parts of secs. 27–34, T. 14 S., R. 27 E. (Dos Cabezas quadrangle), but some deposits are probably located at widely separated points along the south flank adjacent to small gold prospects. The gold-bearing gravels in the gulches in the main mineralized area are thin near the mountains and thicker toward the south near the village of Dos Cabezas (sec. 32, T. 14 S., R. 27 E.). The gold is flat, ragged, and fairly coarse; one report claims that nuggets ranging from 1 to 20 ounces were found.

The placers on the north flank of the Dos Cabezas Mountains (Teviston district) are found in mountain gulches and on pediments at the edge of the mountains. Most of the placer mining was concentrated in the area between Gold Gulch (sec. 24, T. 13 S., R. 26 E., Luzena quadrangle) and Ash Gulch (sec. 22 and 27, T. 13

S., R. 26E.). The placer gravels on the pediment drained by Gold
Gulch consist of coarse to fine granitic sand with some clay and
many coarse, semirounded boulders. Gravels sampled to a depth of
6 feet assayed $4.08 per cubic yard.

Production history: The Dos Cabezas placers reportedly were discovered in 1901, but lode deposits in the district were known in the
1860's and worked intermittently since the 1870's. Although some
reports suggest that the placers were known before 1901, I have
found no production records from that time. Most of the placer
gold was recovered by drywashing the gravels, and, when water was
available, by sluicing and panning.

The placers in the Teviston district have been worked intermittently since the 1900's, but earlier history is unknown. Small dryland
dredges worked placers in Gold Gulch in 1933, and at the Inspiration placers during the period 1937–38. The Ash Gulch placers
were actively worked during the period 1930–31.

Production records combine gold recovery from the Teviston and
Dos Cabezas districts, although the placers in the Teviston district
were richer than those in the Dos Cabezas district.

Source: The gold in the placers was derived from erosion of gold-
bearing quartz veins exposed throughout the Dos Cabezas Mountains.
Most of the important lode-gold mines occur within, and near, a
major fault zone 2 ½ miles north of Dos Cabezas village where
small, closely spaced gold-quartz-sulfide veins occur; other gold
mines are north of this fault zone. A geochronologic study of the
mountain range indicates that some gold-quartz veins are younger
than 29 m.y. (million years).

Literature:

Allen, 1922: Discovery; location; origin (Dos Cabezas district).

Bray, 1933: Describes dryland dredge used at Gold Gulch.

Church, 1887: Notes nonactivity in placer mining, although lode
mining was active.

Engineering and Mining Journal, 1931: Assay results of sampling
at Gold Gulch placer.

Erickson, 1968: Dates mineralized quartz veins.

Gardner and Johnson, 1934: Placer-mining techniques in Gold
Gulch; drywashing; type of gravel.

Heikes and Yale, 1913: Value of gravels; size of large nugget
(Teviston district).

Land, 1931: History; size of nuggests; emphasis on lode deposits
(Dos Cabezas district).

U.S. Bureau of Mines, 1931: Location of placer-mining operation.

————1939–41: Placer-mining operations at Inspiration placers; drag-line dredge; amount of gravel handled; no location for placers.

Wilson, 1961: Location; history; depth of gravels; size of gravels (Teviston district); production.

Wilson, Cunningham, and Butler, 1934: Bedrock geology; history and description of lode mnies; does not describe placers.

2. COURTLAND-GLEESON DISTRICT (TURQUOISE DISTRICT)

Location: Near Maud Hill and south end of Gleeson Ridge in low group of hills east of the south tip of the Dragoon Mountains, T. 19 S., R. 25 E.

Topographic map: Gleeson 15-minute quadrangle.

Geologic map: Gilluly, 1956, Geologic map of parts of the Benson and Pearce quadrangles, Arizona (pl.5), scale 1:62,500.

Access: Gleeson is accessible by light-duty roads leading 11 miles south from U.S. Highway 666 at Pearce, or from a light-duty road leading 8 miles west from U.S. Highway 666 1 mile north of Elfreda, or 16 miles east from U.S. Highway 80 at Tombstone.

Extent: During the 1930's minor gold placers were worked in gulches and on pediments east of Gleeson. Most placer mining was concentrated in an area $\frac{1}{2}$ mile long by $\frac{1}{8}$ mile wide located $1\frac{3}{4}$ miles east of the Gleeson Post Office (sec. 33, T. 19 S., R. 25 E.). The placer is a thin mantle of auriferous gravel and soil on a limestone pediment at the base of Maud Hill. Coarse gold was washed from the gully leading west from the Copper Belle mine (sec. 32, T. 19 S., R. 25 E.) half a mile north of Gleeson.

Production history: The placers were worked in 1910 and again during the period 1932–35. At the base of Maud Hill, tests on 100 cubic yards of placer ground indicated that the gravels averaged $1.12 per cubic yard.

Source: The placer is said to be derived from gold-bearing quartz veins. The literature relating to the Gleeson ore deposits does not describe gold-quartz veins but does describe gold in the lead-silver deposits. Gold was also recovered as a byproduct of copper ores from the Copper Belle mine.

Literature:

Gilluly, 1956: Discusses the age of mineralization in the Courtland-Gleeson district (p. 62, 160).

Wilson, 1927: Discusses mineralization and types of lode mines; no placer description.

————1961: Location; size of gold particles; characteristics of gravels; average gold content per cubic yard; minerals in placer gravels; production.

3. BISBEE-WARREN DISTRICT (GOLD GULCH PLACER)

Location: South flank of the Mule Mountains. Tps. 23 and 24 S., R. 25 E.

Topographic map: Bisbee 15-minute quadrangle.

Geologic map: Hayes and Landis, 1964, Geologic map of the southern part of the Mule Mountains, scale 1:48,000.

Access: A light-duty road leads 2 miles east from Warren to Gold Gulch.

Extent: Minor gold placers are found in the sand and gravel of Gold Gulch, which drains south from the vicinity of Gold Hill in the Mule Mountains (west tier of sections in Tps. 23 and 24 S., R. 25 E.)

Production history: Gold placers were known to occur in Gold Gulch as early as 1902; they were worked sporadically until 1932, then steadily to 1941. In 1934, 27 placer mines recovered 246 ounces of gold, mostly from Gold Gulch.

Source: The small placer gold deposits in Gold Gulch, southeast of the mining towns of Bisbee and Warren, were derived from small gold-bearing silica veins that mineralized parts of the Glance Conglomerate (Cretaceous) during late Cretaceous or early Tertiary time. The mineralizing solutions deposited gold-bearing silica with minor galena, sphalerite, and chalcopyrite along fractures in the conglomerate. The silica veins are not economically important.

Literature:

Ransome, 1904a: Notes presence of placer gold; source; economic importance.

———1904c: Repeats description in 1904a.

Trischka, 1938: Explains origin of placer gold.

U.S. Bureau of Mines, 1935: Reports placer production from Gold Gulch.

Wilson, 1961: Virtually repeats Ransome (1904a); production.

4. HUACHUCA PLACERS (HEREFORD OR HARTFORD DISTRICT)

Location: East flank of the Huachuca Mountains, 3 miles north of the international boundary. Tps. 23 and 24 S., and Rs. 20 and 21 E.

Topographic maps: Hereford and Sunnyside 15-minute quandrangles.

Geologic map: Hayes and Raup, 1968, Geologic map of the Huachuca and Mustang Mountains, scale 1:48,000.

Access: Ash Canyon is accessible by a light-duty road leading east from State Highway 92, 25 miles east of Bisbee.

Extent: Placer gold deposits that also contain some scheelite have been worked in Ash Canyon; most of the gold is found in the bottom of the canyon between the 5,000- and 6,500-foot elevation (secs. 1 and 2, T. 24 S., R. 20 E.; sec. 6, T. 24 S., R. 21 E., Sunnyside quadrangle; secs. 31 and 32, T. 23 S., R. 21 E., Hereford quadrangle). In 1937 gravels consisting of 8 feet of barren overburden and 2 feet of gold-bearing material were mined; the upper 2–3 inches of cemented gravel bedrock was also mined. Gold was recovered from the gravels near the Harper mine (possibly the Harper Tungsten mines in Bear Canyon (approximately sec. 1, T. 24 S., R. 19 E., Sunnyside quadrangle)).

Production history: The placers in the Huachuca Mountains were known during the late 1800's; but they apparently attracted little attention until 1911, when a nugget weighing about 22½ ounces was found, and miners were reportedly recovering $3–$4 per day. No placer production is recorded for 1911, however. In 1919 another large nugget weighing 8½ ounces was recovered from the Old Timer placer in Ash Canyon. The placers have been worked intermittently during the 20th century.

Source: The gold is thought to be derived from auriferous quartz veins found in the granite upstream from the placers; these small veins, which have not been described in the literature, probably contain some scheelite.

Literature:

Blake, 1899: Reports placer mining near the Harper mine.

Gardner and Allsman, 1938: Depth of gravel; thickness of overburden and of pay streak; boulder and clay content; type of bedrock; placer-mining operations with power shovel.

Mining World, 1911: Production per day per man.

U.S. Geological Survey, 1918: Location; placer-mining activity; no recorded production.

————1919: Presence of scheelite in placers; size of nuggets; production.

Weber, 1948: Summary of bedrock geology; no placer description; indicates age of scheelite and gold mineralization.

Wilson, 1941: No placer description; locates Harper tungsten mine.

————1951: Notes that gold placers were worked during depression years; no placer description.

————1961: Location; production; describes type of gravels; distribution of gold in gravels; source of gold.

OTHER DISTRICTS

5. CALIFORNIA DISTRICT

One hundred sixteen ounces of placer gold was recovered in 1906–7; no information has been found relating to the placer occurrence. The California district is located in the vicinity of Paradise (T. 17 S., Rs. 30 and 31 E.) on the east side of the Chiricahua Mountains.

Literature:
  U.S. Geological Survey, 1906: Reports placer production from California district.

6. PEARCE DISTRICT

A silver-gold placer was reported to occur in gravels at the east and west margins of Pearce Hill (secs. 4 and 5, T. 18 S., R. 25 E., Pearce quadrangle). The placers are said to be derived from quartz veins in Pearce Hill that contain silver halides and free gold. No record of placer production is credited to this district by the U.S. Bureau of Mines. Wilson (1961, p. 71) states that placer gold-silver was the first shipment of ore from the district in 1895 and that additional shipments were made during the period 1917–27. The production from placer material containing very high amounts of silver was said to be worth $8,700.

Literature:
  Wilson, 1961: Production; thickness of placer material; gold-silver values in material; source.

7. RUCKER BASIN

One ounce of placer gold was recovered in 1957 from Rucker Basin (T. 19 S., R. 29 E.), 15 miles southwest of the California district on the west flank of the Chiricahua Mountains. No information has been found relating to this placer occurrence.

GILA COUNTY

8. PAYSON (GREEN VALLEY) DISTRICT

Location: South of the Mogollon Rim in the Tonto Basin, T. 10 N., R. 10 E.

Topographic map: Payson 15-minute quadrangle.

Geologic maps:
  Lausen and Wilson, 1925, Geologic map of the Payson district, Arizona (pl. 1), scale 1 inch equals approximately 3 miles.
  Wilson, Moore, and Peirce, 1959, Geologic map of Gila County, scale 1:375,000.

Access: Placers are about half a mile west of State Highway 87, about 5 miles south of Payson, and are accessible by dirt roads leading from the highway to the Ox Bow mine.

Extent: Placer gold was recovered from gravels below the outcrop of the Ox Bow mine (sec. 32, T. 10 N., R. 10 E.). Most of the gold recovered was from surface gravels, but in 1939 some gold was recovered from deeper gravels said to be an old channel not related to the stream channels.

Production history: The Payson (Green Valley) district was first explored during 1875–76; the early prospectors in the 1870's and 1880's were attracted by rich float containing abundant free gold. Although many of these miners were experienced in placer mining, very little placer gold was found in the district. The placers below the Ox Bow vein were worked sporadically for a number of years between 1910 and 1960 during the rainy seasons, but they have not produced much gold.

Source: The gold was derived from the Ox Bow vein and was concentrated in the thin soil on the hillside. The gold particles occur as coarse flat nuggets to a quarter of an inch long and are a deeper color than the vein gold, probably owing to less silver.

Literature:

Lausen and Wilson, 1925: Describes occurrence and character of placer gold.

Mining Journal, 1939b: Reports placer activity in old gravel channel.

Wilson, 1961: Repeats description of Lausen and Wilson (1925).

9. GLOBE-MIAMI DISTRICT

Location: Foothills of the Pinal and Apache Mountains, T. 1 S., R. 15 E.; Tps. 1 and 2 N., Rs. 14 and 15 E. (unsurveyed).

Topographic maps: Globe 15-minute quadrangle; 7 ½-minute quadrangles—Inspiration, Lost Gulch, Gold Gulch, and Pinto Creek; Haunted Canyon (Pinto Creek); Globe (Pinal Creek); Pinal Peak (Gap and Catsclaw Flat; upper Pinal Creek).

Geologic maps:

Peterson, 1960, Geologic map of the Haunted Canyon quadrangle, scale 1:24,000.

Peterson, 1962, Geologic map of the Globe-Miami district, Pinal County, Arizona (pl. 1), scale 1:24,000. (Lost Gulch, Gold Gulch, Pinto Creek, and Pinal Creek)

Peterson, Gilbert, and Quick, 1951, Geologic map of the Castle Dome area, Gila County, Arizona (pl. 1), scale approximately 1:6,000. (Gold Gulch)

Ransome, 1904, Geologic map of the Globe quadrangle, scale 1:62,500. (Pinal Creek; Gap and Catsclaw Flats)

Access: From Globe to Lost Gulch, dirt roads lead about 4 miles west from State Highway 88, 4½ miles north of Globe, to mining areas near Lost Gulch; to Gold Gulch, an improved road leads 2½ miles north from U.S. Highway 60–70, 12 miles west of Globe, to Castle Dome mine area; to Pinto Creek, a dirt road leads 5–6 miles northwest from road to Castle Dome mine to Pinto Creek and Haunted Canyon; to Pinal Creek, Sixshooter Canyon Road, three-quarters of a mile south of Globe, leads 3 miles south to placer area, paralleling Pinal Creek.

Extent: Small placer deposits occur in six localities within the Globe-Miami district. Some placers are near the major copper-mining areas, but they are not necessarily derived from the copper ores.

Lost Gulch, northwest of Globe, is the most productive placer area; the gulch drains southeast from the vicinity of Myberg Basin and the south flanks of Flat Top and Sleeping Beauty Mountains in the Copper Cities mining area to the Inspiration tailing pond (before construction of the pond, Lost Gulch drained to Pinal Creek). The gold placers occur along Lost Gulch and adjoining benches for an undetermined distance, but they were most actively worked in the mile-long part of the creek along the south flank of Sleeping Beauty Mountain (T. 1 N., R. 14 E., Inspiration quadrangle).

Gold was recovered from the gravels near the Golden Eagle vein (unlocated), said to be a short distance north of Miami. This placer area includes a group of gulches that were mined in 1933 and is probably the deposit referred to as the Inspiration placer, located north of Claypool (probably in sec. 16 or 17, T. 1 N., R. 15 E., Globe quadrangle).

Placers occur in the gravels in Gold Gulch in the Castle Dome mining area (west half of T. 1 N., R. 14 E. Inspiration quadrangle). The exact location of the placers is uncertain, but the gold probably was found in gravels half a mile south of the Castle Dome mine.

Placer gold was recovered in small amounts from unlocated placers along Pinto Creek, which heads near Mount Madera in the Pinal Mountains and drains northwestward to the Salt River. One locality where gold probably was recovered is the gravels near the junction of Pinto Creek and Haunted Canyon (Haunted Canyon quadrangle), which were prospected for many years by one man who lived at that locality (Nels P. Peterson, written commun., 1969).

In the southern part of the Globe-Miami area, placer deposits are found in Pinal Creek (Globe and Pinal Peak quadrangles) and in Gap and Catsclaw Flats (Pinal Peak quadrangle). Nels P. Peterson (written commun., 1969) reported that he was told that the gravels were sluiced along the bed of Pinal Creek from about a quarter of a mile above 66 Ranch (SW¼ sec. 13, T. 1 S., R. 15 E.) nearly to the edge of Globe, a distance of about $3\frac{1}{2}$ miles.

The placers adjacent to Pinal Creek at Gap and Catsclaw Flat (sec. 24, T. 1 S., R. 15 E., Pinal Peak quadrangle) occur in an area about 4,000 feet long and 1,500 feet wide east of Pinal Creek near 66 Ranch. Gold was also recovered from gravels in Pinal

Creek on the northeast slope of the Pinal Mountains—near the source of Pinal Creek (Pinal Creek quadrangle). There is no evidence to suggest placer gold accumulations in Pinal Creek downstream from Globe, although the Old Dominion mine and veins system is north of Globe and was once famous for the free gold contained in the copper ores.

Production history: The placers in the Globe-Miami district were worked intermittently from the late 1880's until 1961. Most of the gold was washed from the gravels of Lost Gulch and Pinal Creek. For most years, Lost Gulch was the only placer credited with production; no production directly attributed to Gold Gulch and Pinto Creek has been recorded. The gold recovered from Lost Gulch ranges from fine to fairly coarse; the largest nugget recovered weighed about 2 ounces. The placers were mined by sluicing, rocking, or drywashing; daily returns from the placer areas were low.

Source: The placers have resulted from the erosion of small gold-silver and gold-pyrite veins not associated with the copper porphyry deposits in the Globe-Miami district. In the vicinity of Lost Gulch, numerous small discontinuous pyritic veins occur in the Precambrian schists exposed between Inspiration and the Copper Cities mining area; the placers in this gulch probably were derived from these deposits. The small placers in Gold Gulch probably were derived from the erosion of the Continental vein, which contains low concentrations of gold in the copper ore. The source of the gold in Pinal Creek area is not known, but probably is the small gold-bearing veins in the Precambrian schists exposed in the Pinal Mountains.

Literature:

Blake, 1899: Notes placers in Lost Gulch; source of gold.

Engineering and Mining Journal, 1893: Reports recovery of considerable amounts of coarse gold from Pinto Creek placers.

Hinton, 1878: Lists mines and placers; cites production of placers.

Peterson, 1962: Placer deposits are not described; gold-silver veins that may be a source of the placers are described.

Peterson, Gilbert, and Quick, 1951: Notes placers in Gold Gulch; describes ores that may be the source of the placer gold.

Ransome, 1903: Notes placers in Pinal Creek.

——1904b: Notes placers in Lost Gulch, Gold Gulch, and Pinal Creek; locates placers in Pinal Creek.

Trippel, 1888: Production (Pinal Creek and Lost Gulch); locates placers.

——1889: Production statistics for 1888 (Pinal Creek and Lost Gulch).

U.S. Bureau of Mines, 1934, 1935: Production; names of productive placer claims.

Wilson, 1961: Names five placer areas; location; describes gravels and size of gold particles (Gap and Catsclaw Flats; Lost Gulch); placer mining during the period 1932–33.

10. BARBAROSSA-DRIPPING SPRING PLACERS (BANNER DISTRICT)

Location: Flanks of the Dripping Spring Mountains, T. 3 S., R. 15 E.; T. 4 S., R. 16 E.

Topographic maps: El Capitan and Hayden 7 ½-minute quadrangles; Christmas 15-minute quadrangle.

Geologic maps:
Ransome, 1923a, Geologic map of the Ray quadrangle, scale 1:12,500.
Willden, 1964, Geologic map and sections of the Christmas quadrangle, Arizona (pl. 1), scale 1:62,500.

Access: Jeep trails lead 1–2 miles south to placer area from Dripping Spring Road at a point about 3 miles west of the junction with State Highway 77, 18 miles south of Globe.

Extent: Placers are found on the southwest and northeast flanks of the Dripping Spring Mountains.

The Barbarossa placer, in the old Troy district, is in the SW. cor. sec. 31, T. 3 S., R. 15 E., (Hayden quadrangle) between two forks of Steamboat Wash on the southwest flank of the Dripping Spring Mountains. The deposit consists of soil and loose detritus developed on Troy Quartzite.

The Dripping Spring placers are in the Dripping Spring district near the Cowboy mine on the northeast flank of the Dripping Spring Mountains in the NW. cor. sec. 30, T. 3 S., R. 15 E. (El Capitan quadrangle). These deposits are found in pediment gravels resting on Gila Conglomerate.

Other placers are reported to occur in the southern end of the Dripping Spring Valley north of Christmas in sec. 17, T. 4 S., R. 16 E. (Christmas quadrangle), about 8 miles southeast of the Dripping Spring placers. These deposits (properly in the Banner district) are found in alluvium in gulches that drain northeast to Dripping Spring Valley.

Production history: The Barbarossa placer was discovered in 1907— the recorded production (1907–13) from this placer is credited to the Riverside district on the Southwest flank of the Dripping Spring Mountains (Pinal County) by the U.S. Bureau of Mines. Apparently, only part of the gold recovered from placers on the south flank of the Dripping Spring Mountains has been reported, for an estimate of gold recovery as high as $2,000–$3,000, including one nugget weighing 22 ounces, was made before 1923, and a few prospectors worked the gravels intermittently until recent years.

The placers on the northeast flank of the Dripping Spring Mountains have been worked intermittently for years; production records frequently group the Dripping Spring district and the Banner district, making it difficult to differentiate the two localities. The placers near the Cowboy mine are in gravels 20–30 feet thick and were worked from shafts, tunnels, and underground stopes; about 10 percent of the gold recovered was finer than 100 mesh, but nuggets weighing as much as half an ounce were found. Placer gold worth $3,000 reportedly was recovered from these deposits in 1927, but this amount was not reported to the U.S. Bureau of Mines and is not included in the production table. The placers in the southern part of the Dripping Spring Valley have been worked within the past 30 years and apparently were worked by a dryland dredge in 1940. This ground was known as the Bywater claim; Mr. Bywater also owned the placer ground in the Dripping Spring district.

Source: The Troy district, near the crest of the Dripping Spring Mountains T. 3 S., R. 14 E., was intensely prospected in the early years of this century, but no large commercial ore bodies were found. The free gold in the Barbarossa placer was probably derived from gold localized near the contact between the Devonian Martin Formation and the Cambrian Abrigo Formation (N. G. Banks, oral commun., 1969).

The placers in the Dripping Spring area near the Cowboy mine may have been derived from erosion of the gold-silver vein material of the mine. Free gold was reported from bunches of ore near the surface of the vein.

The origin of the placers farther south in the Dripping Spring Valley is unknown.

Literature:

Mining and Scientific Press, 1907: Barbarossa placer—Reports discovery of placer gold near head of Steamboat Springs Canyon.

Ransome, 1923a: Barbarossa placer—location; production; type of placer gravels.

U.S. Bureau of Mines, 1940: Dripping Spring placer—production; mining operation located.

Wilson, 1961: Barbarossa placer—virtually repeats Ransome (1923); reports elevation of placer; production. Dripping Spring placer—location; type of placer gravels; mining operations (1931–33); size of gold particles; production; source.

OTHER DISTRICTS

11. MAZATZAL MOUNTAINS

Placer gold was recovered between 1910 and 1911 from unknown deposits in the vicinity of Mount Ord and Reno Pass (T. 6 N., Rs. 4–10 E.) in the central Mazatzal Mountains. No information has been found describing gold deposits in this area.

12. YOUNG DISTRICT

Placer gold was recovered from unlocated deposits along Spring Creek, west of Young, in the mesas south of Diamond Butte (probably T. 9 N., R. 12 E.). No information has been found describing ore deposits in this district.

Literature:

Wilson, 1961: States placers recovered in Spring Creek district.

GRAHAM AND GREENLEE COUNTIES

13. CLIFTON-MORENCI DISTRICT

Location: On the west side of the San Francisco River near Clifton in the vicinity of the copper-mining area surrounding Morenci, Tps. 4 and 5 S., Rs. 28 and 29 E.

Topographic map: Clifton 15-minute quadrangle.

Geologic maps:

Lindgren, 1905a, Geologic map and sections of the Clifton-Morenci district (pl. 1), scale 1:62,500 (see also Lindgren, 1905b).

Wilson and Moore, 1958, Geologic map of Graham and Greenlee Counties, scale 1:375,000.

Access: U.S. Highway 666 parallels Chase Creek, west of Clifton, and light-duty roads branching off the main highway lead to Gold Gulch, Chase Creek, and Morenci Gulch placers.

Extent: Small placers were mined in gulches draining the hills surrounding the copper mining area of Morenci. The placers occur in Gold Gulch, Chase Creek, and Morenci Gulch.

Gold Gulch was a noted placer area during the 1870's. The gulch flows about 3 miles to the southwest from the southwest side of the Morenci open pit to Eagle Creek (T. 4 S., Rs. 28 and 29 E.). Gold Gulch is narrow, and the gold is concentrated in a number of bars along the lower course of the creek (sec. 19, T. 4 S., R. 29 E.; sec. 25, T. 4 S. R. 28 E.).

Chase Gulch flows southeast between Copper King Mountain and Morenci to Clifton. Placer mining was concentrated between the Old Rock House and the town of Clifton (secs. 23–25, T. 4 S., R. 29 E.). The gold was recovered from tributary gulches and from elevated gravels resting on Gila Conglomerate above the present level of Chase Creek.

Fine flakes of gold were recovered from Morenci Gulch, a south-east-flowing tributary of the San Francisco River. Part of Morenci Gulch is now covered by the Morenci Tailing Pond; the undisturbed part of the gulch is in sec. 12, T. 5 S., R. 29 E.

Production history: It is difficult to isolate the recorded production from the different creeks in the Clifton-Morenci district from production statistics that include the production for all the placers in Graham and Greenlee Counties. Discovery of placer deposits in the Clifton-Morenci district in 1870 by ranchers from Silver City, N. Mex., stimulated mining interest in the area and copper mining began soon afterward. Frederick Remington, a famous western artist, lived and placered for a year or so at Gold Gulch; Remington is said to have uncovered $6,000 in placer gold in 3 weeks by removing boulders and rocks that covered a depression in which the rich gravel had settled.

During the 20th century placer mining has been done on a small scale in Gold Gulch and Chase Creek; the daily earnings per man working the Chase Creek placers in 1933 was frequently less than 50 cents.

Source: The placer gold was derived from oxidized gold-bearing veins associated with the intrusive porphyry and commonly found along or near the porphyry-sediment contact. In Gold Gulch, these veins are narrow and irregular and occur mostly along both sides of Pinkard Gulch and the lower part of Gold Gulch; they were prospected on a small scale during the late 1800's and produced some gold from isolated pockets.

On the ridge between Chase Creek and Morenci Canyon, gold-bearing fissure veins occur associated with a porphyry sill between quartzite and limestone; at the Hormeyer mine (sec. 22, T. 4 S., R. 29 E.) the vein contained a little copper, much lead carbonate, and native gold.

Literature:

Allen, 1922: Quotes Lingren (1905a).

Dinsmore, 1911b: Early history of placer mining in Gold Gulch.

Lindgren, 1905a: Notes presence and location of placers; describes gold veins that are probable source of placer gold.

———1905b: Describes gold veins in Gold Gulch; brief history of placer mining.

Mining Reporter, 1906: Notes past placer mining in many gulches; describes development in lode-gold mining.

Moolick and Durek, 1966: Placer discovery.

Tovote, 1910: Source of placer gold.

Wilson, 1961: Virtually repeats Lindgren (1905a, b). Placer-mining operations in the early 1930's.

14. SAN FRANCISCO RIVER PLACERS

Location: Along the San Francisco River from Dorsey Gulch south to the junction with the Gila River, Tps. 4 and 5 S., Rs. 29 and 30 E.

Topographic maps: Clifton and Guthrie 15-minute quadrangles.

Geologic maps:

Lindgren, 1905a, Geologic map and sections of the Clifton-Morenci district (pl. 1), scale 1:62,500 (see also Lindgren, 1905b).

Wilson and Moore, 1958, Geologic map of Graham and Greenlee Counties, scale 1:375,000.

Access: Light-duty road parallels San Francisco River near and north of Clifton. Dirt roads lead from U.S. Highway 666 to various points along the lower San Francisco River south of Clifton.

Extent: The placer deposits that occur along the course of the San Francisco River are logically divided into two groups. In the part of the river north of Clifton, the gold is found in ancient river gravels 50–60 feet above the level of the present riverbed, from the vicinity of Dorsey Gulch (sec. 5, T. 4 S., R. 30 E., Clifton quadrangle) south to the vicinity of Clifton (sec. 30, T. 4 S., R. 30 E.) The placers in benches of the river near Oroville (sec. 7, T. 4 S., R. 30 E.) have attracted the most attention. Here, the richest gold-bearing layers occur in thin pay streaks in channels at or near bedrock on the benches above the riverbed. The Bokares placer, 4 miles north of Clifton at Evans Point (sec. 6 or 7, T. 4 S., R. 30 E.), was also actively mined.

South of Clifton, the river is curved and flows between bluffs of hard Gila Conglomerate. Gold is contained in old river gravels resting on the conglomerate in some of the curves. The Smuggler placer mine is at a bend in the San Francisco River in sec. 14, T. 5 S., R. 29 E. (Guthrie quadrangle); the gravels here contained much fine gold. Placers mined by the "Frisco Placer Mining Company" in the early 1900's are probably located in sec. 31, T. 5 S., R. 29 E. (Guthrie quadrangle), 8 miles downstream from Clifton. The gravels at this placer deposit are 3–12 feet thick.

Production history: The placers were discovered in the 1870's and were actively prospected and developed during the 1880's. At that time, much money was spent to develop pipelines and water supplies to mine the gravels above Clifton by the hydraulic technique; the operations were not a financial success, although a fair amount of gold was recovered. The deposits were mined more or less continuously until the 1940's; small-scale mining techniques were used, the men frequently tunneling through the gravels to reach the richest parts near bedrock.

The production statistics here are, as for the Clifton-Morenci district, difficult to isolate from those of other placer districts in Graham and Greenlee Counties; but the bulk of the recorded production for the counties was derived from the placers along the San Francisco River.

Source: The placer gold in the San Francisco River above Clifton was derived from small gold-bearing veins, associated with porphyry dikes, that crop out in Dorsey and Colorado Gulches (for example, Colorado, Black Prince, and Golden Eagle veins) and were once mined for gold. The gold in the river south of Clifton was probably derived from these veins and from the small gold veins around Morenci that were the source of gold in the Clifton-Morenci placers.

Literature:

Allen, 1922: Quotes Lindgren (1905a).

Blake, 1899: Location; problems connected with large-scale placer-mining operations.

Burchard, 1882: History of discovery; placer operations in 1881.

———1883: Placer-mining developments; production.

———1884: Location; describes depth of placer gravels.

Dinsmore, 1911b: Operations of Frisco Placer Mining Co.; history; depth of gravels; value of placer gravels.

Hamilton, 1884: Repeated from Burchard (1884).

Lindgren, 1905a: Failure of past hydraulic operation; source of placers.

———1905b: Location, age, and source of placer gravels.

Mining Journal, 1938b: Placer-mining operations at Bokares and Smuggler placer.

Mining Reporter, 1906: Location; past mining operations.

Trippel, 1888: Production for 1887.

Wilson, 1961: Location; characteristics of gravels; size of gold particles; placer operations in 1933.

15. GILA RIVER PLACERS

Location: Along the Gila River from the junction of the San Francisco River southwest to Bonita and Spring Creeks. Tps. 5 and 6 S., Rs. 28 and 29 E.

Topographic map: Guthrie 15-minute quadrangle (Graham and Greenlee Counties).

Geologic maps:

Heindl and McCullough, 1961, Geologic map and sections showing location of infiltration gallery, mills, and springs, lower Bonita Creek area, Graham County, Ariz. (pl. 1), scale 1:68,000.

Wilson and Moore, 1958, Geologic map of Graham and Greenlee Counties, scale 1:375,000.

Access: Placers are accessible by dirt road that parallels the north bank of the Gila River. The road is 1½ miles north of Safford.

Extent: Placers are found along the Gila River below the junction of the San Francisco River (Greenlee County) to the mouth of Spring Creek (Graham County). Some of the gravels of the Gila Conglomerate between the San Francisco River and Eagle Creek (Greenlee County) contain fine flakes of gold. Gold is found in ancient river gravels that mantle terraced bluffs of Gila Conglomerate along the Gila River downstream from the mouth of Eagle Creek to Bonita Creek (Graham County); the gold ranges in size from flakes up to wiry particles a quarter of an inch long. About 10–12 miles downstream from Eagle Creek, the Gila River makes a wide bend between Bonita and Spring Creeks. At this location (approximately sec. 20 and 21, T. 6 S., R. 28 E.) an alluvial flat was tested for placer gold at the property known as the Neel placer.

Production history: Placer production from the Gila River has been very minor compared with production from the San Francisco River. Placer production recorded for Graham County from 1907 to 1910 was recovered from the area separated into Greenlee County in 1910. Placer production from the Gila River in Graham County is given under the Lone Star district.

Some of the deposits along the Gila River between Eagle Creek and Bonita Creek are said to carry gold values of 15–50 cents per cubic yard. Tests of the Neel placer made in 1933 and 1938 indicate that the gravels averaged 60 cents per cubic yard. Actual production from the property was small.

Source: The origin of the gold in the gravels is unknown, but it was probably derived from gold eroded from the Clifton-Morenci district and transported by the San Francisco River to the Gila River. Small gold veins in the Gila Mountains may have contributed some gold to these minor placers.

Literature:

Allen, 1922: Quotes Lindgren (1905a).

Lindgren, 1905a: Notes placer occurrence in lower San Francisco River and Eagle Creek; size of gold.

————1905b: Virtually the same description as 1905a.

Mining Journal, 1938d: Placer-mining operations at Neel Placer; gold values per cubic yard.

Wilson, 1961: Location; characteristics of gravels; size of gold particles; gold values per cubic yard. Placer mining in 1933.

OTHER DISTRICTS

16. RATTLESNAKE DISTRICT

Five ounces of gold was recovered in 1932 from an undescribed placer in this district, located on the north flank of the Galiuro Mountains (Tps. 8 and 9 S., Rs. 19 and 20 E.) in southern Graham County.

MARICOPA COUNTY

17. BIG HORN DISTRICT

Location: Big Horn Mountains southeast of the Harquahala Mountains, Tps. 3–5 N., Rs. 8 and 9 W., in particular, T. 5 N., R. 9 W.

Topographic maps: Aquila 15-minute and Big Horn Mountains 15-minute quadrangles.

Geologic map: Wilson, Moore, and Peirce, 1957, Geologic map of Maricopa County, scale 1:375,000.

Access: From Aquila it is about 10 miles south on light-duty road to the Big Horn Mountains. Dirt roads lead to many areas within the mountain range.

Extent: Although placers within the Big Horn Mountains have produced notable amounts of placer gold since 1900, very little is known about the location and character of the placer deposits. The U.S. Bureau of Mines Mineral Yearbooks name five placer claims in the district: Big Horn, Sweeney, Tiger, Borian, and Davenport. None of these placer claims are accurately located. Most lode mines in the area are in T. 5 N., R. 9 W., and it seems reasonable to suggest that some placer deposits might be in that general vicinity.

Production history: Placers were worked in the Big Horn district from 1933 to 1942; most of the production seems to have been recovered by individuals working several placer claims.

Source: The source of the placer gold is unknown, but the placers probably originated by erosion of gold-bearing veins in the Big Horn Mountains.

Literature:

U.S. Bureau of Mines, 1934–41: Placer production; names of placer claims given in the years 1934 (for 1933), 1935 (for 1934), 1940 (for 1939), 1940 (review of 1940).

Wilson, Cunningham, and Butler, 1934: Describes lode mines and general geology.

18. VULTURE DISTRICT

Location: South flank of the Vulture Mountains in the vicinity of the Vulture mine, Tps. 5 and 6 N., Rs. 5 and 6 W.

Topographic map: Vulture Mountains 15-minute quadrangle.

Geologic map: Wilson, Moore, and Peirce, 1957, Geologic map of Maricopa County, scale 1:375,000.

Access: From Wickenburg it is 14 miles south on Vulture mine road to mine and placers in the immediate vicinity.

Extent: The only description I have found of the placer deposits in this district is given by Wilson. The overshadowing importance of the Vulture mine is certainly the explanation for the lack of detail about the placer ground, which covers an area about 3 miles square in Red Top Basin (secs. 24 and 25, T. 6 N., R. 6 W.) and extends for a distance of 2 miles southeast of the Vulture mine in Vulture Wash (secs. 6 and 7, T. 5 N., R. 5 W.). Red Top Basin is a pediment formed on Precambrian schist and mantled by gold-bearing gravels generally less than 10 feet thick. The gold is coarse and angular and is generally concentrated on bedrock.

Production history: The placers were apparently worked from the time of discovery of the Vulture mine. In the early days of the district (from 1867 to 1880), many large nuggets weighing about $\frac{1}{2}$ to 1 ounce were recovered, and reportedly, some weighing 5 ounces. Although most of the richest gravels were worked out by 1880, small-scale drywashing in the area continued from that time until 1948.

Source: Small gold-bearing quartz veins in the immediate vicinity are thought to be the source of the placer gold in Red Top Basin. The origin of the gold in Vulture Wash is considered to be partly the Vulture vein and partly other small gold veins.

Literature:

Wilson, 1961: Location; history; past production; character of placer gravels; size of nuggets; origin of placers.

19. SAN DOMINGO DISTRICT

Location: South flank of the Wickenburg Mountains, northeast of the Hassayampa River, Tps. 6 and 7 N., Rs. 3 and 4 W.

Topographic maps: Wickenburg and Red Picacho 7½-minute quadrangles.

Geologic map: Wilson, Moore, and Peirce, 1957, Geologic map of Maricopa County, scale 1:375,000.

Access: Jeep trails lead to low hills on north side of U.S. Highway 60–70–89 near San Domingo Wash, 7 miles southeast of Wickenburg, 3 miles northwest of Morristown.

Extent: The placers in the San Domingo district are found in San Domingo Wash, its tributaries and adjacent washes, and on mesas between gulches. The placer area is southwest of a low range of hills that includes the prominent San Domingo Peak; the drainage of the area is southwestward to the Hassayampa River. Many washes are mentioned in the literature as scenes of active placer mining, but the topographic maps of the area locate only San Domingo Wash, which drains through secs. 25 and 26 and 35, T. 7 N., R. 4 W., and secs. 2 and 3, T. 6 N., R. 4 W. (Wickenburg quadrangle) and Little San Domingo Wash (Tps. 6 and 7 N., R. 3 W., Red Picacho quadrangle). Other washes placered are Old Woman Gulch, Rogers Wash, American Gulch, Spring Gulch, and Sanger Gulch. Old Woman Gulch is described as a southern tributary of San Domingo Wash; two tributaries enter San Domingo Wash from the south in the placer area—one tributary joins San Domingo Wash in sec. 2, T. 6 N., R. 4 W., the other, in sec. 5, T. 7 N., R. 4 W. Rogers Wash, which has been described by reporters in a series of articles (Carter, 1911, 1912; Dinsmore, 1911a; Hafer, 1911), is $2\frac{1}{2}$ miles long and is probably the wash located $1\frac{1}{2}$ miles northwest of the mouth of San Domingo Wash. The principal placer area in Rogers Wash is probably in sec. 26, T. 7 N., R. 4 W. (Wickenburg quadrangle). The Alibu placer (SE¼ sec. 28 T. 7 N., R. 4 W.) is adjacent to the highway midway between this wash and Monarch Wash on the north. "Spring Gulch" is probably "Tub Spring Gulch," a headward tributary to San Domingo Wash (T. 7 N., R. 3 W.). The other two placer-bearing washes are not located. The gravels in the Hassayampa River contain gold for a few miles below San Domingo Wash.

Production history: The San Domingo district has the largest recorded placer gold production in Maricopa County and produced continuously (although on a small scale) from 1905 to 1951. During the early 1960's gold was recovered as a byproduct of gravel operations and by large-scale operations of a mobile dryland dredge.

It is not known when the placers were discovered, but the district was actively worked during the period 1870–80. It is said that the greatest production occurred during this time, when individual placer miners recovered $15–$100 per day. Old Woman Gulch was a large producer in 1875 (one report indicates 1885), and American Gulch reportedly produced "fortunes" for California miners at the same time.

Between 1910 and 1912, the district was actively prospected by Mr. John Sanger, who started the Lotowana Mining Co. and planned to mine 4,000 acres of placer ground on a large scale. The company concentrated its exploration efforts in Rogers Wash, where the average value of the ground was 68 cents per cubic yard, but where the value of some strips was as high as $1.19 to $1.36 per cubic yard. A dam built by Mr. Sanger across San Domingo Wash failed because the intended reservoir filled with sand and gravel before operations had finished one season. Actual production by the Loto-wana Mining Co. is not known.

Since 1912 the records indicate only small-scale operations in the district until the late 1950's. In 1959 the MacDonald Construction Co. recovered gold from gravels that were sorted according to size. Fine gravel (minus $3/16$ in. mesh) contained most of the gold; coarser gravel (plus $3/16$ in. to minus $3/4$ in.) was sent to a nugget trap part of the time. During the period 1960–62 a dryland dredge, "Geraldine," owned by United Placers Industries, Inc., worked the gravels in the San Domingo district. The operation of this dredge received considerable attention in mining journals; but production records are confidential, and the success of the operation is not known.

Source: The San Domingo district is predominantly a placer-mining district, and there is very little literature available that describes the gold-bearing veins. The origin of the placer gold is said to be from Precambrian and post-Cambrian veins in the area.

Literature:

Allen, 1922: Location; source; virtually repeats information de-scribed by Carter (1912).

Arizona Engineer and Scientist, 1961: Describes dryland dredge; average gold values of placer deposits; mining techniques; size of gold-bearing gravels.

Carter, 1911: Location; describes Rogers Wash; character of gravels; fineness and shape of gold; accessory minerals; distribution of gold in gravels; placer-mining techniques and development.

——1912: Drywashing along the Hassayampa River. Locates Rogers Wash; bedrock geology; distribution of gold in gravels; placer-mining techniques and development.

Dinsmore, 1911a: Location; placer-mining developments; history of placer mining; early production from placer washes; character of placer gravel; grade of gravel.

Engineering and Mining Journal, 1961: Describes dryland dredge "Geraldine"; mining techniques.

Hafer, 1911: Brief description of placer ground.

Heikes and Yale, 1913: Describes Rogers Wash; virtually repeats articles by Carter; adds no new information.

Jahns, 1952: Describes pediment gravels; no placer description.

Roseveare, 1961: Placer-mining operations during the period 1959–61; size of gold-bearing gravel fractions.

Wilson, 1961: Location; history; source of gold; size of nuggets; dis-tribution of gold in gravels; gold values in gravels.

20. CAVE CREEK DISTRICT

Location: East flank of the New River Mountains, north of Paradise Valley, Tps. 6 and 7 N., R. 4 E.. Lower Cave Creek, south of Cactus, T. 3 N., R. 3 E.

Topographic maps: All 7½-minute quadrangles—Cave Creek, New River Mesa, Sunnyslope.

Geologic map: Wilson, Moore, and Peirce, 1957, Geologic map of Maricopa County, scale 1:375,000.

Access: From Phoenix, 42 miles northeast on light-duty road to the town of Cave Creek.

Extent: Very little information other than production records has been found relating to the placers in the Cave Creek district. Cave Creek heads near Cramm Mountain in the New River Mountains (T. 7 N., R. 4 E., unsurveyed) and flows southwest to Deer Valley, north of Phoenix. The placers are probably located along Cave Creek in the vicinity of the Maricopa and Phoenix gold mines (secs. 8 and 9, T. 6 N., R. 4 E., New River Mesa quadrangle). The only description of the placers I have found is that written by McConnell (1911), who describes a new placer discovery made in 1910. The ground, which was not located except for the district, contains gold-bearing gravels underlying soil and cemented gravels. The placers are found from 1 to 16 feet deep between the cement layer and bedrock; tests indicate that the ground values average 85 cents to $5.10 per cubic yard.

Other placers are apparently found in the Winifred district on Cave Creek about 12 miles south of the Cave Creek district. U.S. Bureau of Land Management survey plats locate placer claims in sec. 22, T. 3 N., R. 3 E. (Sunnyslope quadrangle).

Production history: The earliest record of placer production from the Cave Creek district is for 1888; placers may have been worked before that time. Small amounts of placer gold were recovered by transitory miners between 1908 and 1915, and 1934 and 1941. From 1939 to 1941, the small production of the Cave Creek district was combined with production from the Camp Creek district in the record.

The Winifred district produced a small amount of placer gold in 1932 (amount was not given).

Source: The paucity of information about the Cave Creek district precludes definite conclusions regarding the origin of the placer gold. The Maricopa and Phoenix gold mines, the largest lode-gold mines in the area, are in quartz veins in altered Precambrian schist. These deposits, and possibly other small gold veins, could have been the source of the placers in the Cave Creek district.

The small production of placer gold from the Winifred district may have been derived from the veins found at the Avelina, Divide, Corona, and La Fe lode claims.

Literature:

McConnell, 1911: Placer ground discovered; thickness and depth of placer gravels; average gold value per cubic yard.

Trippel, 1889: Production statistics for 1888.

Wilson, Cunningham, and Butler, 1934: Describes lode mines and general geology.

OTHER DISTRICTS

21. AQUA FRIA DISTRICT

Placer gold was recovered from Moores Gulch (T. 8 N., Rs. 2 and 3 E.), a tributary to the Aqua Fria River, north of Lake Pleasant. Placer gold was recovered from other areas along the river, but no descriptions of any deposits have been found.

22. DADS CREEK

Placer gold was recovered in 1935 and reported to the U.S. Bureau of Mines. The location of this creek is unknown.

Literature:

U.S. Bureau of Mines, 1935: Gives placer production.

23. EAGLE TAIL MOUNTAINS

In 1912 a small amount of placer gold was recovered from deposits in the Santa Rosa group of claims during assessment work. The Eagle Tail Mountains are in western Maricopa County and extend westward into Yuma County (Tps. 1 and 2 N., Rs. 10 and 11 W.). I have not found the location of the Santa Rosa group.

Literature:

U.S. Geological Survey, 1912: Reports placer production.

24. NEW RIVER DISTRICT

Placer gold was recovered from this district in 1933. The district is on the south slope of the New River Mountains (Tps. 6 and 7 N., Rs. 2 and 3 E.) but probably also includes minor deposits located in the low hills south of the New River Mountains and north of Deer Valley. The Relief mine (secs. 3 and 4, T. 4 N., R. 1 E.), north of Peoria and west of the New River, was known as a placer before 1917, but only the lode-gold production record was found. The location of the placer deposit worked in 1933 is unknown.

Literature:

Elsing and Heineman, 1936; Gives lode-gold production.

Schrader, Stone, and Sanford, 1917: Notes placer occurrence at Relief mine.

25. PIKES PEAK DISTRICT

The Pikes Peak district is at the eastern end of the Heiroglyphic Mountains, on the west side of the Aqua Fria River (T. 6 N., Rs. 1 and 2 W.). Placer gold was recovered from Morgan Wash in this district during the period 1939–41 and in 1948. The district is known for iron deposits. I have found no descriptions of gold ores.

Literature:

U.S. Bureau of Mines, 1939–41, 1948: Production of placer gold.

26. SUNFLOWER DISTRICT

Sunflower is on the west slope of the Mazatzal Mountains, north of Mount Ord (sec. 13, T. 7 N., R. 8 E.: Reno Pass 7½-minute quadrangle). Placer gold was recovered from this area in 1940, but the location of the deposit is unknown.

MOHAVE COUNTY

27. CHEMEHUEVIS DISTRICT (GOLD WING DISTRICT)

Location: Chemehuevis Mountains (also known as the Mohave Mountains) east of the Colorado River, Tps. 14 and 15 N., Rs. 19 and 20 W.

Topographic maps: Topock 15-minute quadrangle; Needles 2-degree sheet, Army Map Service.

Geologic map: Wilson and Moore, 1959a, Geologic map of Mohave County, scale 1:375,000.

Access: State Highway 95, 9 miles east of Topock, leads south about 15 miles to the vicinity of the Chemehuevis Mountains, about 10 miles north of Lake Havasu City. Dirt roads lead to various placer areas.

Extent: Placer deposits have been worked at many localities in the Chemehuevis Mountains, but the deposits cannot be accurately located on topographic maps of the area because mapping of the Buck Mountain quadrangle is incomplete.

From the description of the deposits, it would seem that most of the placer-mining activity was concentrated on the southwestern flank of the mountains in the area east of the Colorado River. One of the deposits, known as the Calizona placer channel, is 1 mile wide and 3–5 miles long and trends northwest toward the Colorado River. The auriferous gravel is found in a bed of conglomerate or fanglomerate 10–30 feet thick, probably located in or near secs. 26–28, T. 15 N., R. 20 W. (Topock quadrangle). Placers were mined in the Mexican or Spanish diggings in the vicinity of the Red Hills approximately T. 14 N., R. 20 W., Topock quadrangle). In the early 1930's placers were mined on the northeastern side of the mountains in Dutch and Printers Gulches (approximately T. 15 N., R. 9 W., Needles quadrangle).

Production history: The Chemehuevis placers reportedly were discovered in 1857 and have been worked on a small scale since the early 1860's, when many miners drywashed the gravels in the Calizona placer on the southwest side of the mountains. Many of these early miners are said to have recovered as much as $500 per day. Tests of the Calizona placer ground made in 1909, indicated a working average of $3.40 per cubic yard.

Gold was recovered from the Fisher Diggings, Silver Creek, the "49," Chief, and Prentice Gulch properties; the location of these claims is unknown. Tests at the Fisher Diggings made during 1932, reportedly indicated an average of $1.15 per cubic yard.

Source: The lode deposits of the Chemehuevis district are found in quartz veins in Precambrian schist. These veins only locally contain high gold values associated with pyrite and galena, and, apparently, are relatively unoxidized near the surface. The placers are presumably derived from these veins.

Literature:

Blake, 1899: Notes presence of placer gold.

Hedburg, 1909: Location; extent and depth of placer gravels; placer-mining operations; results of sampling; average value of placer ground; size of gold particles.

Mining Journal, 1932c: Placer-mining operations; average grade of gravel at the Fisher diggings.

Moore, 1969: Date of placer discovery.

Randolph, 1901: Notes presence of placer gold.

Wilson, 1961: Location; type of placer gravels; placer operations during the period 1932–33.

Wilson, Cunningham, and Butler, 1934: Describes lode mines in the Chemehuevis district which may be source of placer gold.

28. SAN FRANCISCO DISTRICT (OATMAN DISTRICT)

Location: West flank of the Black Mountains, Tps. 19 and 20 N., Rs. 20 and 21 W.

Topographic map: Oatman 7½-minute quadrangle.

Geologic maps:

Lausen, 1931, Geologic map of the Oatman district, Mohave County, Arizona (pl. 1), scale ≈1:40,000.

Ransome, 1923b, Geologic map of the Oatman district, Arizona (pl. 1), scale 1:48,000.

Wilson and Moore, 1959a, Geologic map of Mohave County, scale 1:375,000.

Access: From Kingman, about 30 miles west to Oatman on the Kingman-Oatman-Topock road; a light-duty road parallels Silver Creek wash west of Oatman, and dirt roads lead to small placer areas near Oatman.

Extent: Small placer deposits have been worked in the vicinity of some gold mines near Oatman and in the valley of Silver Creek, which drains northwest from the Oatman Camp. Three small placers were worked in the Oatman area: (1) Placer gold was recovered from small side streams tributary to Silver Creek in the vicinity of Mount Hardy (secs. 5 and 6, T. 19 N., R. 20 W.). The gold here was of two different colors—pale yellow and darker yellow—and might not have been recovered from the same stream. The size of the gold from this locality varied from small flat flakes to particles as large as the size of wheat grains. (2) Placer gold was recovered from different points along a small stream below the Pioneer mine (sec. 2, T. 19 N., R. 20 W.) The gold here was mostly fine particles and occurred within 3 feet of bedrock. (3) Placer gold was recovered from gravels below the Moss Vein (sec. 19, T. 20 N., R. 20 W.), 2 miles north of Silver Creek. According to Lausen (1931, p. 89), one important ore shoot on the Moss veins contains coarse gold, and he suggests that the placer gold was coarse.

In addition to the small amounts of placer gold recovered in the vicinity of the Oatman Camp, gold was recovered from the valley of Silver Creek, about 5 miles northwest of Oatman (possibly from the gravels found in secs. 31 and 32, T. 20 N., R. 20 W.). This deposit was tested in 1923 and during the period 1932–33, and, although one report indicates that gold values in the 5 feet above bedrock were high and that the average value might be $1 per cubic yard, the amount of gold in the gravels was apparently not sufficient to encourage mining operations. The gold is found in the gravel which overlies an irregular pediment formed on volcanic rocks.

Production history: The richest lode-gold mining area in Mohave County is the San Francisco district. Placer-gold production from this district has been negligible compared with lode production, despite the fact that placers have been worked since about 1865. Although the placers in Silver Creek were prospected and large sums of money expended to investigate the gold-bearing gravels, very little production was recorded. Apparently, most of the placer mining in the district was done on a very small scale.

Source: The small amount of placer gold found in the district was derived from the gold-quartz-calcite veins in Tertiary igneous rocks that formed in late Tertiary time. Lausen (1931, p. 88–89) suggests several factors that might be responsible for the lack of rich placer deposits. The most important is the small size of the gold in the ores; these small particles could be transported for a long distance during floods. Owing to lack of water, the deposits were mined by drywashing machines, which cannot collect fine gold.

Literature:

Allen, 1922: Notes placer occurrence along Silver Creek.

Arizona Mining Journal, 1924: Reports discovery of buried gravels averaging $1.50 per yard (with gold valued at $20.67 per oz).

Doman, 1922: Notes placer-mining activity about 1865.

Lausen, 1931: Locates three placers in Oatman area; size of gold particles; source; reasons for general absence of placers in important lode district.

Mining Journal, 1932b: Reports production from gravels below Warner Gulch.

Ransome, 1923b: Placer-mining operations in Silver Creek.

Salt Lake Mining Review, 1923: Placer-mining developments in Silver Creek; depth of gold-bearing gravels; thickness of rich layer; average value of gravels.

Wilson, 1961: Placer-mining operations during the period 1932–33 in Silver Creek; geologic occurrence of gravels; fineness of gold.

29. KINGMAN AREA PLACERS (McCONNICO AND MAYNARD DISTRICTS)

Location: Northeast flank of the Hualapai Mountains on Kingman Mesa; northwest flank of the Hualapai Mountains. T. 20 N., R. 17 W.; Tps. 20 and 21 N., R. 16 W.

Topographic maps: Kingman 7½-minute quadrangle; Kingman and Williams 2-degree sheets, Army Map Service.

Geologic map: Wilson and Moore, 1959a, Geologic map of Mohave County, Arizona, scale 1:375,000.

Access: From Kingman, 3 miles south on U.S. Highway 66 to Lewis placer area; 6 miles southeast from Kingman on dirt roads to Lookout placer area.

Extent: Three minor placer deposits occur in gravels in the low hills near Kingman. The Lewis placer and the Boulder Creek placer are in the McConnico mining district southwest of Kingman; the Lookout placer is in the Maynard mining district southeast of Kingman.

The Lewis placer is on the property of the Bi-Metal gold mine (sec. 4, T. 20 N., R. 17 W.). The placer gold is found in some small gullies within, or at the border of, the mineralized granite on Kingman Mesa. Some nuggets worth 50 cents were recovered from the area.

The Boulder Creek placer is near the Boulder Creek group of lode claims (approximately sec. 10, T. 20 N., R. 17 W.); the gold veins were located in 1906 after tracing detrital gold to the outcrops.

The Lookout placer is at the north end of the Hualapai Mountains (Tps. 20 and 21 N., R. 16 W.). The placers are found in areas of shallow gulch and hillside gravels, but the exact location is uncertain.

Production history: Most of the placer gold known to have been recovered from the Kingman area placers was recovered from the Lewis placer between 1932 and 1933. A. E. Lewis reportedly recovered $6 to $10 per day at the Bi-Metal property using two sluices. Wilson (1961, p. 34) states that $150 in placer gold was recovered from the Lookout placer during the 1932–33 season, but this recovery was apparently not reported to the U.S. Bureau of Mines.

Source: The Lewis placer resulted from the erosion of free gold concentrated in cavities created by the oxidation of pyrite in the granite at the Bi-Metal property. The Boulder Creek placer resulted from the erosion of parts of the Boulder Creek group of veins. The origin of the Lookout placer is unknown.

Literature:

Engineering and Mining Journal, 1933b: Production in 1933; location of placer ground.

———1933d: Results of sampling Lewis placer.

Schrader, 1909: Lewis placer—location; size of gold particles; source. Boulder Creek placer—location; source.

Wilson, 1961: Location; production for 1932–33.

30. COLORADO RIVER PLACERS

Location: Along the Colorado River from the mouth of the Grand Canyon south to Topock. Tps. 22, 27, 29, N., R. 22 W.; T. 31 N., R. 19 W. (projected).

Topographic maps: All 15-minute quadrangles—Virgin Basin, Black Canyon, Mount Perkins, Spirit Mountain.

Geologic maps:

Longwell, 1936, Area near Colorado River between Black Canyon and the head of Boulder Canyon (pl. 2), scale 1 in. = about 1¾ miles; area near Colorado River between Virgin-Detrital Valleys and head of Iceberg Canyon (pl. 3), scale 1 in. = about 1¾ miles; area near Virgin River south of St. Thomas (pl. 4), scale 1 in. = about 1¾ miles.

Longwell, 1963, Geologic map and sections of area along Colorado River between Lake Mead and Davis Dam, Arizona and Nevada (pl. 1), scale 1:125,000.

Wilson and Moore, 1959a, Geologic map of Mohave County, scale 1:375,000.

Access: From Kingman, the placers along the Colorado River can be reached by the following routes: To the Temple Bar area—42 miles northwest on U.S. Highway 93 to improved road which leads north about 24 miles to Temple Bar. To the Willow Beach area—56 miles northwest on U.S. Highway 93 to Willow Beach road; an improved road leads about 4 miles west to Willow Beach placer area. To the Eldorado Canyon area—46 miles north on U.S. Highway 93 to a dirt road that leads 12 miles west to the vicinity of the placer area opposite Eldorado Canyon. To the Pyramid Rock area—30 miles west on State Highway 68 to Davis Dam; dirt road leads north from Davis Dam about 7 miles to the placer area.

Extent: Placer gold has been recovered from many locations along the Colorado River. Four locations have been described in the literature, and placer gold was probably recovered from other localities as well. Placer gold was found at Temple Bar on the Colorado River north of the White Hills (T. 31 N., R. 19 W., projected; Virgin Basin quadrangle). The gravel containing fine gold was apparently found on both the Arizona and the Nevada side of the river; the gravel bar was inundated by the waters of Lake Mead.

Coarse gold was said to be found at Willow Beach near an outer bow of the Colorado River (T. 29 N., R. 22 W., Black Canyon quadrangle). The bar covers an area of about 250 square feet and rests upon an irregular surface of gneissic granite.

Sand bars opposite Eldorado Canyon on the Arizona side of the Colorado River (T. 27 N., R. 22 W., Mount Perkins quadrangle) contain finely divided gold.

Some moderately coarse gold was recovered from a bench near the river about $2\frac{1}{2}$ miles north of Pyramid Rock. Apparently, this locality now lies beneath Lake Mohave about 4 miles north of Davis Dam (T. 22 N., R. 22 W., Spirit Mountain quadrangle).

Production history: Recorded production from the Colorado River placers is very small. In 1895 a large hydraulic plant was constructed at Temple Bar, but the enterprise quickly met with failure because of the high cost of transporting materials to the area. Placer gold was recovered from this placer in 1935 before inundation of the riverbed by Lake Mead. According to Wilson (1961, p. 34–35), the placer at Willow Beach was worked before 1900, in 1920, and in 1931, but I have found no production record. In 1909 a suction-type dredge was installed to work the gravels opposite Eldorado Canyon, but the dredge failed to extract the fine gold on the first try and subsequently was destroyed during the high waters in the spring of 1910. No production record was found for the locality near Pyramid Rock.

Source: The source of the placer gold at these localities is unknown, but the occurrence of gold-bearing lodes in the surrounding region suggests that much of the gold may be locally derived.

Literature:

Allen, 1922: Placer-mining operations during the period 1909–10 opposite Eldorado Canyon.

Blake, 1899: Temple Bar placers—operation; grade of gravel; source.

Lausen, 1931: Pyramid Rock placer—notes presence of gold northwest of the Catherine district.

Randolph, 1903: Failure of placer-mining operations at Temple Bar.

U.S. Geological Survey, 1910: Failure of placer-mining operations opposite Eldorado Canyon.

Wilson, 1961: Willow Beach placer—location; extent; type of bedrock; size of gold; source; placer-mining operations; production. Eldorado Canyon placer—failure of placer operations. Pyramid Rock placer—location; size of gold.

31. GOLD BASIN AND LOST BASIN DISTRICT

Location: On the east flank of the While Hills and on the east and west flanks of the Lost Basin Range, south of Lake Mead, Tps. 29 and 30 N., Rs. 17 and 18 W.

Topographic map: Garnet Mountain 15-minute quadrangle.

Geologic map: Wilson and Moore, 1959a. Geologic map of Mohave County, Arizona, scale 1:375,000.

Access: From Kingman, 29 miles north on U.S. Highway 93 to Pierce Ferry Road; this road leads northeast about 22 miles to the Gold Basin area and 30 miles to the Lost Basin area.

Extent: The placers in the Gold Basin and Lost Basin districts are found in three major areas: the east and west flanks of the Lost Basin Range and the detrital fan in Gold Basin on the east flank of the White Hills. The placers on the east flank of the Lost Basin Range are found in arroyos incised in bajada gravels of late Miocene and early Pliocene age which cover an area of 8–10 square miles. Many individuals have drywashed the placers at various localities. Five major placer claims are located along this flank of the range. These are the Robeson and Joy lease (sec. 14, T. 30 N., R. 17 W.), the Queen Tut placer (secs. 27 and 34, T. 30 N., R. 17 W.), the Golden Nugget placer (near the intersection of secs. 33 and 34, T. 30 N., R. 17 W.; secs. 3 and 4, T. 29 N., R. 17 W.), the King Tut placer (sec. 9, T. 29 N., R. 17 W.), and the Lone Jack placer (sec. 16, T. 29 N., R. 17 W.). The King Tut placer was the most actively mined placer in the area, and the east flank of the Lost Basin Range is frequently called the King Tut placer area.

On the west flank of the Lost Basin Range, small-scale mining of placers found in Quaternary alluvial fans is still active. These placers occupy an area comparable in size to the placer ground on the east side of the range and are located in the eastern rows of T. 29 and 30 N., R. 17 W.

The gold-bearing gravels of the Gold Basin district are found in arroyos and gulches on the large detrital fan that slopes eastward from the White Hills to Hualapai Wash and is traversed by White Elephant Wash and its tributaries. The Searles placer mine is in sec. 29, T. 29 N., R. 18 W.

Production history: The placers in the Gold Basin and Lost Basin districts were first actively mined in 1931, about 60 years after the discovery of lode gold. Placer gold was recovered from the Gold Basin district in 1909, but, probably because of the isolation of the district, apparently no further placer mining was done until 1931. Placer-mining activity since the early 1930's has been almost continuous but on a small scale. A few mining operations have used power shovels and small dry-concentrating plants to mine the gravels, but most activity was with the small portable drywasher so prevalent in the Southwest.

Owing to the relatively late development of the placers, early miners were able to sample virtually untouched placer ground in this area. On the east flank of the Lost Basin Range, the richer gold-bearing gravels are generally less than 2 feet thick and rest on caliche-cemented gravels. The gold contained in these surface gravels ranges in size from fine dust to nuggets as much as three-quarters of an ounce; in 1941, a nugget valued at $140 (4 oz) was recovered from a placer near the King Tut. Extensive sampling of one section at the King Tut placers in 1933 indicated that the average value of the placers was $1.17 per cubic yard.

The gold-bearing gravels of the Gold Basin area are 1–3 feet thick and rest on cemented gravels. The gold ranges in size from fine to coarse; fragments found range in value from 5 cents to $3.50. The gold is reported to be erratically distributed; thin streaks that yield more than $1.00 per cubic yard are found, but most arroyo gravels contain less than $1.00 per cubic yard.

Source: Recent work by the U.S. Geological Survey indicates that there are many small gold-quartz-carbonate-sulfide veins in the Precambrian rocks. Gold derived from some of these veins is the probable source of the placers.

Literature:

Blacet, 1969: Brief summary of type of lode deposits; notes presence of scheelite in the placers.

Engineering and Mining Journal, 1933b: Placer-mining activity in 1933; location of placer area sampled.

———1941: Size of nuggets.

Mining Journal, 1933: Placer-mining operations; average value of placer gravels at King Tut placer.

U.S. Geological Survey, 1968: Extent and age of placers; average value of placers; source of gold.

Wilson, 1961: Location; placer-mining operation during the period 1932–33; extent and thickness of gravels, accessory minerals; average value of gravels; size of gold particles; source.

OTHER DISTRICTS

32. COTTONWOOD DISTRICT

Placer gold has been recovered intermittently from small deposits in Wright Creek on the northeast side of the Cottonwood Mountains (Tps. 22 and 23 N., Rs. 11 and 12 W.).

Literature:

Wilson, 1961: Reports presence of placer gold in Wright Creek.

33. OWENS (McCRACKEN) DISTRICT

Small amounts of placer gold have been recovered from unlocated placers in this district, located in the McCracken Mountains (T. 13 N., Rs. 14 and 15 W.). No description of the placer gold occurrence has been found.

34. WALLAPAI DISTRICT

Less than 1 ounce of placer gold was recovered from this district, located on the west side of the Cerbat Mountains. The district is famous for rich lead-zinc ores with silver and gold as byproducts. Gold occurs in galena and sphalerite in moderately oxidized ores.

Literature:

Koschmann and Bergendahl, 1968: Describes lode deposits.

PIMA COUNTY

35. ALDER CANYON PLACERS

Location: Northeast slope of Santa Catalina Mountains, T. 11 S., Rs. 16 and 17 E.

Topographic map: Bellota Ranch 15-minute quadrangle.

Geologic map: Wilson, Moore, and O'Haire, 1960, Geologic map of Pima and Santa Cruz Counties, scale 1:375,000.

Access: Alder Canyon, about 10 miles southeast of Oracle, is accessible by dirt roads leading from town along the flank of the mountains.

Extent: Coarse placer gold is found in dissected bars and benches along Alder Canyon and on spurs between tributary gulches from near the Coronado National Forest boundary (sec. 13, T. 11 S., R. 16 E.) to within a few miles of the San Pedro River (east margin of T. 11 S., R. 17 E.).

Production history: The placers were reportedly worked intermittently for many years, but many miners were transient and stayed only a short while, having recovered very little gold. Recorded production is small and for some years included with production from the Old Hat district.

Source: Unknown.

Literature:

Wilson, 1961: Location, extent, size of gold; placer-mining activity during the period 1932–33; production.

U.S. Bureau of Mines, 1935: Gives production for 1934 from Alder Canyon credited to Old Hat district.

36. GREATERVILLE DISTRICT

Location: East flank of the Santa Rita Mountains, T. 19 S., Rs. 15 and 16 E.

Topographic maps: All 15-minute quadrangles—Sahuarita, Empire Mountains, Mount Wrightson, and Elgin.

Geologic maps:

Drewes, 1971a, Geologic map of the Mount Wrightson quadrangle, scale 1:48,000.

———1971b, Geologic map of the Sahuarita quadrangle, Pima County, scale 1:48,000.

Hill, 1910, Sketch map of the Greaterville, Arizona, placer camp.

Access: The Greaterville area is accessible by roads that lead 5 miles west from State Highway 83, about 8 miles north of Sonoita on the junction with State Highway 82.

Extent: The placers in the Greaterville district are found in streams that drain easterly from the Melendrez Pass area in the Santa Rita Mountains to the Cienega Valley. The deposits are in the southeastern part of T. 19 S., R. 15 E., and in the southwestern part of T. 19 S., R. 16 E. The gold-bearing gulches are, from north to south: Empire, Chispa, Colorado, Los Pozos, Hughes, Ophir, Nigger, and St. Louis Gulches, tributaries to Hughes; Louisiana, Graham, Sucker, Harshaw, Kentucky, and Boston. Placers were found not only in the gulch gravels but also in gravels on the hillsides and ridgetops between gulches. Hill (1910) describes the distribution of the gold-bearing gravels in each gulch in detail, and, as his report is well known, I will only summarize his description.

In general, the gold is found in the lower 2 feet of angular gravel overlying bedrock and underlying less rich gravels; in places, the gold was concentrated in natural riffles in the sedimentary bedrock. The gold recovered ranged in size from flakes to large nuggets. Hill (1910, p. 20) states that the gold washed in 1909 ranged from small flakes to particles 0.1 inch in greatest dimension. Most of the largest nuggets were recovered during the early mining period in the district; at that time (1874–86) nuggets worth $1 to $5 (about $1/20$–$1/4$ oz) were common, and one nugget weighing 37 ounces was found.

Production history: The early production is not accurately known. For gold recovered before 1900, estimates range from as high as $7 million to as low as $500,000. During the 20th century the placers have been worked continually by many individuals using rockers. Much of the placer ground has been reworked several times, but a considerable amount of gold is said to remain in the gravels. Various attempts have been made to mine the gravels in different gulches using hydraulic or dredge mining methods, but thickness of overburden made large-scale mining unprofitable. In 1948 a dragline shovel and dryland washing plant treated 90,000 cubic yards of gravel from Louisana Gulch, recovering 535 ounces of gold, an average of 21 cents per cubic yard (or 0.006 oz per cubic yard). This was the largest amount of gold recorded from the placers in any one year during the 20th century.

Source: The placer gold was derived from erosion of free gold-bearing veins genetically related to a quartz latite porphyry (dated at 55.7 ±1.9 m.y.) intrusive into Cretaceous sedimentary rocks. These veins are found near the heads of the gulches and have been mined for the gold content at the Yuba, St. Louis, and Quebec mines.

Literature:

Allen, 1922: Virtually repeats information described by Hill (1910). Adds information about placer-mining operations in 1914.

Black, 1890: Past placer-mining activity.

Blake, 1898: Describes late Pleistocene bison remains found in placer deposit.

———1899: Detailed description of placers. Includes location; names of placer gulches; size of nuggets; placer-mining operations; source of placer gold; repeats part of earlier description by Black (1890). Describes separately size of nuggets found in the "Smith district" now known as Greaterville.

Burchard, 1882: Placer-mining activity at Hughes Gulch is noted.

———1884: History; production.

———1885: History; size of nuggets; placer-mining operations; techniques; production.

Drewes, 1970: Describes mineralization in district; distribution of gold.

Gardner and Johnson, 1935: Placer-mining techniques by drift mining.

Heikes and Yale, 1913: Thickness of placer gravels; gold values per cubic yard; fineness of gold; production from 1903 to 1912.

Hill, 1910: Detailed description; extent, thickness, and character of gravels; size and fineness of gold; bedrock geology; placer-mining techniques, operations; production.

Hinton, 1878: Notes placer occurrence; size of large nugget.

Koschmann and Bergendahl, 1968: Location; source; production.

Maynard, 1907: Detailed description of sampling. Grades of gravel given for each gulch.

Randolph, 1901: Production estimates.

Raymond, 1875: Reports placer ground on the east side of the Santa Rita Mountains; names prospectors who made discoveries.

———1877: Repeats information of newspaper article describing new discovery of placer gold at Greaterville; size of nuggets; methods of mining district; refers to as "Smith district"; does not locate district; names miners.

Root, 1915: Placer mining in 1914; average grade of gravel; depth of water "sufficient for dredging purposes."

Schrader, 1915: Virtually repeats Hill's (1910) description.

Wilson, 1961: Placer-mining activity during the period 1874–1948; history; quotes Schrader (1915); production; source.

37. ARIVACA DISTRICT (LAS GUIJAS PLACERS)

Location: The Las Guijas and San Luis Mountains, Tps. 20 and 21 S., Rs. 9 and 10 E.

Topographic map: Arivaca 15-minute quadrangle.

Geologic map: Wilson, Moore, and O'Haire, 1960, Geologic map of Pima and Santa Cruz Counties, scale 1:375,000.

Access: From Arivaca Junction on U.S. Highway 89, it is 23 miles west on light-duty road to Arivaca. Many roads lead to placer areas in the surrounding mountains.

Extent: Placer gold has been found in most of the gravels that mantle the flanks of the Las Guijas Mountains and in many gulches that head in the mountains. Most of the placer mining was concentrated on the northeast flank of the range, south of Las Guijas Creek; here, mesa gravels between the edge of the mountains and the creekbed are notably gold bearing for a length of 2½ miles and a width of 1 mile (sec. 25, T. 20 S., R. 9 E., secs. 30–32, T. 20 S., R. 10 E.). Durzano and Pesqueria Gulches (sec. 32, T. 20 S., R. 10 E.; sec. 5, T. 21 S., R. 10 E.) also contain placer gold.

Placers are found in the gravels in large basins and wide arroyos on the southern slope of the Las Guijas Mountains, northeast of Arivaca Wash (NE¼ and NW¼ T. 21 S., Rs. 9 and 10 E.). Apparently the gold in these gravels, in contrast to the placers on the north side of the mountains, is irregularly distributed.

Farther south, in San Luis Canyon (vicinity of the SE. cor. T. 21 S., R. 9 E.), placers are reported in some interarroyo bench gravels on the dissected pediment in that area.

Production history: The placers in the Arivaca district have been worked since, and possibly before, the 1850's. The occurrence of placer gold in the region was so well known that the mountains were named "Guijas," which means rubble or conglomerate in which placer gold is usually found. The early production of placer gold from the area is unknown, but it has been estimated to be about $150,000. In the 1850's Ignacio Pesqueria and a band of followers defeated in a revolution in Mexico fled to the Las Guijas area, where they obtained sufficient gold from the gravels to finance another revolution that in 1856 enabled Pesqueria to become Governor of the State of Sonora, Mexico.

Since that time, placer mining has continued in the district with varying degrees of intensity. In 1905 the New Venture Placer Mining Co. made plans to mine the gravels by hydraulic techniques after sampling that indicated an average of $1.69 per cubic yard on tests of 4,800 cubic yards but apparently did not begin operations. The same company mined the placers in Durzano, Pesqueria, and Yaqui Gulches in 1915 with a Clark agitating sluice, which uses small amounts of water. Later, in 1933, the pediment and gulch gravels at the northern foot of the Las Guijas Mountains were worked on a large scale; the gold-bearing gravels contained much black sand and some cinnabar. Placer wolframite was recovered from gravels in the low range of hills north of Las Guijas Creek.

Source: The geology of the Arivaca district is poorly known. Many prospects and mining claims are located in the Las Guijas Mountains, and the placer gold was undoubtedly derived from local gold-bearing veins, the occurrence and nature of which is unknown.

Literature:

Allen, 1922: Location; extent of placer gravel; source; thickness of gravel; distribution, size, and shape of gold particles; placer-mining operations; gold values per cubic yard.

Bryan, 1925: Definition of "Guijas."

Trippel, 1888: Notes placer-mining activity.

Willis, 1915: Location; history; production estimates; thickness and extent of placer gravels; character of gold particles; grades of gravel; distribution of gold in gravel.

Wilson, 1941: Reports wolframite placers.

———1961: Location; history, placer-mining activity during the period 1932–33; source.

38. SIERRITA MOUNTAINS PLACERS (PAPAGO AND PIMA DISTRICTS)

Location: Southeast and southwest flanks of the Sierrita Mountains, T. 18 S., Rs. 10 and 12 E.

Topographic maps: Palo Alto Ranch and Twin Buttes 15-minute quadrangles.

Geologic map: Wilson, Moore, and O'Haire, 1960, Geologic map of Pima and Santa Cruz Counties, scale 1:375,000.

Access: From Sahuarita on U.S. Interstate 19, it is about 15 miles west on improved roads to Sierrita Mountains. Dirt roads lead to placers on the flanks of the mountains.

Extent: Placer gold is found in two areas in the Sierrita Mountains— Ash Creek in the Papago district and Armargosa Arroyo in the Pima district.

Ash Creek drains the southwest flank of the Sierrita Mountains; the placer is located in the vicinity of the Sunshine-Sunrise group of mining claims (sec. 12, T. 18 S., R. 10 E.; sec. 7, T. 18 S., R. 11 E.; Palo Alto Ranch quadrangle). Placers were also found in Pascola Canyon, not identified on the topographic map.

The placer in Armargosa Arroyo and tributaries in the Tinaja Hills (southeast flank of the Sierrita Mountains) is in secs. 20, 21, 28, and 29, T. 18 S., R. 12 E. (Twin Buttes quadrangle). The topographic map shows "Tinaja Wash" in this area. The gold was recovered from gravels in the arroyo and its tributaries and from the thin soil on the hillside.

Production history: The recorded production from these placers is small, but the placers are said to have been worked before the discovery of gold in California and reportedly produced considerable amounts of gold before records were kept. Elsing and Heineman (1936, p. 98) attribute a placer production of $250,000 to the Papago district placers; this sum may be an estimate of the early production of the area.

Source: Unknown.

Literature:

Allen, 1922: Location; placer-mining operations.

Browne, 1868: History; early date of placer mining.

Elsing and Heineman, 1936: Placer-production estimate for Papago placers.

Wilson, 1961: Location, production for Papago district. Location, placer-mining activity; accessory minerals in placers for Pima district.

39. BABOQUIVARI DISTRICT

Location: East and west flanks of the Baboquivari Mountains. T. 18 S., R. 7 E.; T. 20 S., R. 8 E.

Topographic maps: Presumido Peak and Baboquivari Peak 15-minute quadrangles.

Geologic map: Wilson, Moore, and O'Haire, 1960, Geologic map of Pima and Santa Cruz Counties, scale 1:375,000.

Access: State Highway 286 parallels the east flank of the Baboquivari Mountains, and service roads lead about 8 miles west from the highway to mining areas.

Extent: Placer gold is found about 5 or 6 miles southeast of Baboquivari Peak at the eastern foot of the mountains. The gold-bearing gravels range in thickness from 6 to 11 feet and are found in benches and bars along a large east-trending wash (Placeritos Wash or Shaffer Wash, T. 20 S., R. 8 E., Presumido Peak quadrangle); the gravels contain many boulders and some clay.

The Fresnal placer claims are near the Lost Horse lode claims in the Baboquivari Mountains; the location of these claims is not known to me, but the name suggests Fresnal Canyon on the west side of the mountains near the Allison lode-gold mine (T. 18 S., R. 7 E., projected, Baboquivari Peak quadrangle).

Production history: The placers in the Baboquivari district were discovered during, or shortly before, the early 1930's; since that time, production has been small, and gold recovery was reported for only 4 years (1935, 1938, 1940, 1958). During 1933, the Edna J. Gold Placer Mines, Inc., leased 680 acres in the placer area on the east side of the mountains; the company planned to work a bar of gravel that contained about 50,000 cubic yards of gravel reportedly averaging 65 cents per cubic yard. In 1938 another company held the Fresnal placer claims. Apparently, there was very little development of these placer claims, as there is no record of any large-scale mining operations in the area.

Source: Unknown.

Literature:

Mining Journal, 1938a: Placer-mining developments at the Fresnal Placers.

Wilson, 1961: Placers at eastern side of Baboquivari Mountains—location, extent; placer-mining activity; average grade of gravels, depth of gravels.

40. CABABI (COMOBABI) DISTRICT

Location: Southeastern side of the South Comobabi Mountains. T. 17 S., R. 5 E. (projected; on Papago Indian Reservation).

Topographic map: Sells 15-minute quadrangle.

Geologic map: Wilson, Moore, and O'Haire, 1960, Geologic map of Pima and Santa Cruz Counties, scale 1:375,000.

Access: The main mining area and placers are about 4 miles north

of State Highway 86, 5 miles east of Sells, the Papago Indian Reservation Headquarters. Dirt roads lead from the highway north
to the mining area.

Extent: Placer deposits of unknown extent are in the vicinity of the
Jaeger gold mine (approximately sec. 3, T. 17 S., R. 5 E.). The
Jaeger mine is on an undulating pediment formed on shales at the
eastern margin of the south side of the South Comobabi Mountains.

Production history: Placer gold has been produced sporadically from
this area since 1911. Apparently, most of the work was done by
Mexican and Indian miners, who sold the gold to the merchants
at Sells (formerly called Indian Oasis, a name once applied to the
Cababi district by the U.S. Geological Survey) from 1911 to 1915.

Source: Gold-bearing quartz veins occur in the southeastern part of
the South Comobabi Mountains. Gold occurs associated with iron
oxides in the oxidized parts of the veins. At the Jaeger mine, several
narrow gold-quartz veins form the lode; similar veins probably were
the source of the placer gold.

Literature:

Mining Journal, 1946: Reports location and development of Jaeger
lode claims and prospecting of adjacent placer claims.

U.S. Geological Survey, 1914–15: Reports placer mining in "Indian
Oasis" district.

Wilson, Cunningham, and Butler, 1934: Describes lode mines in
district.

41. QUIJOTOA DISTRICT

Location: East and west flanks of the Quijotoa Mountains. Tps. 15
and 16 S., R. 2 E. (projected; on Papago Indian Reservation).

Topographic map: Quijotoa Mountains 15-minute quadrangle.

Geologic map: Wilson, Moore, and O'Haire, 1960, Geologic map of
Pima and Santa Cruz Counties, Arizona, scale 1:375,000.

Access: From Tucson, 82 miles west on State Highway 86 to Quijotoa;
dirt roads lead into the mountains.

Extent: The placers in the Quijotoa Mountains apparently are widely
distributed, for some reports indicate placers as far south as the
international boundary. The deposits have been mined by Papago
Indians and Mexicans since 1774, but information regarding the
locations of the deposits mined between 1774 and 1849 is lost. Since
1880 most of the placer-mining activity apparently has been concentrated in the area around Quijotoa, Covered Wells (Maish
Vaya), and Pozo Blanco (Stoa Vaya). The specific deposits described
in the literature are difficult to locate because the area is within the
Papago Indian Reservation and sections are not surveyed and gulch
and claim names given to the placer claims by miners have not been

retained. Placers were recovered from gravels in the area known as Horseshoe Basin that surrounds the old town of Quijotoa (3 miles west of State Highway 86 and 3 miles south of Covered Wells), from two gulches (Homestake and Midas) in the vicinity of Covered Wells (Maish Vaya), and from an area located 3 miles south of Pozo Blanco (Stoa Vaya) and 1 mile west of the mountains.

Production history: The Quijotoa placers have been worked on a small scale from 1774 to the present. The early production is unknown; since the 1860's many reports indicate that $3,000–$7,000 per year in placer gold was recovered by Papago Indians and Mexican miners. The gold was recovered from the unconsolidated surface gravels and from the underlying caliche-cemented gravel. The gravels are said to average more than 80 cents per yard. In 1910 a Quinner pulverizing machine and a Stebbins separator table was used in Horseshoe Basin to recover gold from the gravels, but most of the mining was done by individuals who pulverized the gravels by hand before using drywashers or bateas to separate the gold.

Source: Numerous deposits of vein gold are in the Quijotoa Mountains; erosion of these veins has concentrated the gold in various placers along the flanks of the mountains.

Literature:

Allen, 1922: Location; character of gravels; distribution of gold in gravels; placer-mining techniques; placer-mining operations during the period 1905–6 and in 1910.

Blake, 1899: Location; production; general history of placer mining.

Browne, 1868: Notes long activity at Quijotoa placers.

Bryan, 1925: Locates Pozo Blanco and Horseshoe.

Burchard, 1885: Placer discovery in 1884; location; extent; length and thickness of auriferous gravels; production.

Elliott, 1884: History; early date of placer-mining activity; extent of placer ground; Quijotoa.

Fickett, 1911: Placer-mining techniques; distribution of gold in gravels and caliche.

Heikes and Yale, 1913: Location; placer-mining operations; character of placer gravels; production from 1903 to 12.

Hinton, 1878: Notes placer occurrence; size of gold.

Mining Journal, 1939c: Reports average value of placer gravel at Mackey Brothers claims; depth of gravels.

——1940: Reports nuggets as large as half an ounce in weight.

Randolph, 1903: Notes placer-mining activity.

Stephens, 1884: History; early placer mining (1774–1849); placer-mining activity during the period 1883–84.

Wilson, 1961: Location; history; placer-mining activity (1906, 1910, 1932–33); distribution of placer gold at Pozo Blanco and Horseshoe Basin; average value of placer gravels; source.

42. AJO DISTRICT

Location: Adjacent to the New Cornelia mine in the Little Ajo Mountains, T. 12 S., R. 6 W.

Topographic map: Ajo 15-minute quadrangle.

Geologic map: Gilluly, 1946, Geologic map and sections of the Ajo mining district, Pima County (pl. 20), scale 1:12,000.

Access: From Yuma, 117 miles east on U.S. Interstate 8 to Gila Bend; from there 42 miles south on State Highway 85 to Ajo.

Extent: A small placer deposit is found in gravels in Cornelia Arroyo (secs. 23 and 26, T. 12 S., R. 6 W.).

Production history: 24 ounces of gold was recovered during the period 1932–33 at a time when the New Cornelia copper mine was closed; apparently, no placer gold has been recovered since that time.

Source: The gold is evidently derived from the oxidized part of the New Cornelia ore body, which contains about 0.0067 ounce of gold per unit of copper.

Literature:

Gilluly, 1946: Location; source; production.

OTHER DISTRICTS

43. EMPIRE DISTRICT

The Empire district is in the Empire Mountains (T. 18 S., R. 17 E.), which extend northeastward from the Santa Rita Mountains; the lode deposits of the district are primarily base-metal replacement and contact deposits. In 1935, 2 ounces of placer gold was reportedly recovered, but the location and source of this placer is unknown.

Literature:

Wilson, 1951: Describes principal mines in the Empire district.

44. OLD BALDY DISTRICT (MADERA CANYON PLACERS)

Madera Canyon is on the northwest slope of the Santa Rita Mountains (T. 19 S., R. 14 E.). Reportedly there was considerable placer mining in the lower part of the Madera Canyon alluvial fan in the late 1880's, but I have found no record of any placer gold produced in this district. The only recent activity reported was for 1932–33, when the deposits were sampled.

Literature:

Schrader, 1915: Location, extent, depth of gravels; early placer-mining activity.

Wilson, 1961: Virtually quotes Schrader (1915); placer-mining activity during the period 1932–33.

45. SILVER BELL DISTRICT

No information has been found describing placer deposits in the Silver Bell district, Silver Bell Mountains (Tps. 11 and 12, S., R. 8 E.) about 40 miles northwest of Tucson.

PINAL COUNTY

46. OLD HAT DISTRICT (CAÑADA DEL ORO PLACERS; SOUTHERN BELLE PLACERS)

Location: Northwest and northeast flanks of the Santa Catalina Mountains, T. 10 S., Rs. 14–16 E.

Topographic maps: Mammoth and Oracle 15-minute quadrangles.

Geologic maps:

Creasey, 1967, Geologic map and sections of the Mammoth quadrangle, Pinal County, Arizona (pl. 1), scale 1:48,000.

Wilson and Moore, 1959b, Geologic map of Pinal County, Arizona, scale 1:375,000.

Access: From Oracle, dirt roads lead southeast and southwest to the placer areas on both sides of the mountains.

Extent: The Cañada del Oro placer area is the large alluvial fan at the northwest end of the Santa Catalina Mountains north and west of Samaniego and Oracle Ridges (T. 10 S., Rs. 14 and 15 E., Oracle quadrangle). Gold is found in gravel beds that reportedly range in thickness from 6 feet at the creek side to 252 feet at the top of the alluvial fan. The gold-bearing gravel occurs over a wide strip along the creek and in adjacent hillsides and extends south into Pima County.

The Southern Belle placer is in creek beds below the Southern Belle mine on the northeast flank of the Santa Catalina Mountains (unsurveyed secs. 19 and 20, T. 10 S., R. 16 E., Mammoth quadrangle). The gold is said to be concentrated in a pay streak on top of red clay material derived from the decomposition of diorite (diabase of Creasey, 1967).

Production history: The placers in Cañada del Oro have been known and worked for many years. Some reports state that Spaniards may have worked the gravels in the early 1700's, but I have found no estimates of the gold recovered by these early miners. The deposits have been worked on a small scale throughout most of the 20th century; small amounts of gold were recovered in most years. Most of the gold occurred as well-rounded particles ranging in size from grains worth a few cents to one-fourth-ounce nuggets. During the early 1930's a nugget worth $25 (at $20.67 per oz) was recovered, and a 16-pound lump containing about 40-percent quartz was said to have been found in the late 1800's.

The placers near the Southern Belle mine were known in 1884 and may have been worked before that time. These deposits have no recorded production and were probably worked out in the early 1900's. Reportedly, the gold recovered was coarse, as some large nuggets were said to have been taken out.

Source: According to Wilson (1961, p. 61–62,) gold-bearing veins in the upper reaches of Cañada del Oro, which heads on the north flank of Mount Lemmon, were the probable source of the placer gold in that area. I have found no detailed information about these veins, which were worked at the Copeland, Kerr, Matas, and other prospects.

The gold in the Southern Belle placer is presumably derived from the gold-bearing Southern Belle ore deposit.

Literature:

Allen, 1922: Quotes Heikes and Yale (1913).

Blake, 1899: Notes presence of placer gold; source at Southern Belle.

Browne, 1868: Placer-mining activity at Cañada del Oro.

Burchard, 1885: Placer-mining activity at Southern Belle.

Burgess, 1903: Locates Southern Belle placer; average gold content in 50-pound sample; describes lode mine.

Creasey, 1967: Describes geology and ore deposits in vicinity of Southern Belle mine (p. 82–83). Does not describe placers.

Heikes and Yale, 1913: Location; thickness of gravels; character and origin of placer gravels; gold values per cubic yard; size of nuggets; production from 1903 to 1912; placer-mining operations.

Hinton, 1878: Notes placer occurrence; profitable workings.

Hodge, 1877: Notes history of mining activity.

Wilson, 1961: History; placer-mining activity during the period 1932–33; origin of gold. Quotes Heikes (1913).

OTHER DISTRICTS

47. CASA GRANDE DISTRICT

The Casa Grande district includes the Silver Reef and Slate Mountains in southwestern Pinal County. Free gold has been found in veins at the Mammon mine, western Slate Mountains (T. 10 S., R. 4 E.), and probably occurs in other parts of the range. Ellis locates gold placers in the alluvial valley on the east side of the Slate Mountains (T. 10 S., R. 6 E.). No information has been found about the placer gold recovered from this district in 1922.

Literature:

Ellis, 1962: Map locates placers in vicinity of Slate Mountains.

Wilson, Cunningham, and Butler, 1934: Describes ores at Mammon mine.

48. GOLDFIELD DISTRICT

Small amounts of placer gold have been recovered from deposits in the northwestern part of the Superstition Mountains (T. 1 N., R. 8 E.). The gold was probably derived from free gold in oxidized quartz veins in granite, which were mined at the Morman Stope of the Young mine.

Literature:

Wilson, Cunningham, and Butler, 1934: Describes ores in the district.

49. MINERAL CREEK DISTRICT

In 1888, placer gold valued at $400 was recovered from Mineral Creek (Ray district) on the west flank of the Dripping Spring Mountains (T. 3 S., R. 13 E.) in the eastern part of Pinal County.

Literature:

Trippel, 1889: Production.

50. PIONEER (SUPERIOR) DISTRICT

Placer gold was recovered from the southwestern part of the Pioneer district in the vicinity of Picket Post Mountain. The probable source of the gold was eroded material derived from oxidized veins in the area.

SANTA CRUZ COUNTY

51. ORO BLANCO DISTRICT

Location: Oro Blanco Mountains north of the United States-Mexico boundary, T. 23 S., Rs. 10 and 11 E.

Topographic maps: Ruby and Oro Blanco 15-minute quadrangles.

Geologic map: Wilson, Moore, and O'Haire, 1960, Geologic map of Pima and Santa Cruz Counties, scale 1:375,000.

Access: From Nogales, about 25 miles northwest on State Highway 289 to Oro Blanco. Dirt roads and jeep trails lead south to placer areas.

Extent: The Oro Blanco Mountains are named for the fact that the placer gold found in the area is so alloyed with silver that it is silvery white. Placer gold is said to occur in almost every ravine and gulch, on many hillsides, and on surfaces where the soil is reddish from decomposed pyrite. Alamo Gulch and its neighboring gulches (T. 23 S., R. 10 E., Oro Blanco quadrangle) reportedly contained the richest placers; gold recovered from Alamo Gulch was valued at \$16 per ounce (at \$20.67 per oz). Placers found at the mouth of Warsaw Gulch on California Gulch (also called Oro Blanco Viejo Gulch; sec. 29, T. 23 S., R. 11 E., Ruby quadrangle) contained gold valued at \$10 per ounce (at \$20.67 per oz). Most of the gold ranged in size from flour to small nuggets.

Production history: During the past century, most of the placer mining was done with crude implements and with an inadequate water supply. During the 20th century some large-scale placer mining has been attempted; most of these operations were unsuccessful, however, because of the fine size of the gold.

Source: The placers were derived from the numerous gold-silver-bearing veins in the Oro Blanco Mountains.

Literature:

Bird, 1916: Notes placer-mining activity.

Blake, 1899: Location; extent of placer ground; character of placer gravels; placer-mining techniques.

Girand, 1933: Placer-mining techniques; activity in 1933.

Koschmann and Bergendahl, 1968: Placer production from 1896 to 1904.

Randolph, 1903: Placer-mining activity in 1902.

Wilson, 1961: Location; extent; areas of placer concentration; history; placer-mining activity during the period 1906–32; early production estimates.

52. NOGALES DISTRICT

Location: In the vicinity of Mount Benedict southwest of the Santa Cruz River, T. 23 S., R. 14 E.

Topographic map: Nogales 15-minute quadrangle.

Geologic map: Wilson, Moore, and O'Haire, 1960, Geologic map of Pima and Santa Cruz Counties, scale 1:375,000.

Access: From Nogales, about 2 miles northeast on State Highway 82 to dirt road leading northwest about 2 miles to mines on the south slope of Mount Benedict.

Extent: Schrader (1915, p. 355), reported that placers occur in Guebabi Canyon, which drains the northwest flanks of the Patagonia Mountains and crosses an alluvial plain to the Santa Cruz River.

Placers occur on the east side of Mount Benedict near the Santa Cruz River (Bird, 1916, p. 10). This area is about 1–3 miles southeast of the mouth of Guebabi Canyon.

Production history: The recorded production of placer gold from the Nogales district is very small, and the exact location of the placers worked is unknown, although Schrader (1915) suggests that the placers in the Guebabi canyon were among the oldest and largest placer mines in the area. F. S. Simons (oral commun., 1971) reports that no evidence was seen to indicate the presence of placer gravels and that the occurrence of gold in the canyon is unlikely.

Source: The placers found on the east side of Mount Benedict were probably derived from gold-bearing veins known to occur in this isolated mountain of Precambrian granite.

Literature:

Bird, 1916: Notes presence of placer gold on slopes of Mount Benedict near the Santa Cruz River.

Mining Review, 1910b: Reports high concentration of placer gold in gravels surrounding Nogales; an exaggerated account.

Schrader, 1915: Location of placers in Guebabi Canyon; extent of gravels; early placer-mining activity.

Wilson, 1961: Quotes Schrader.

53. PATAGONIA DISTRICT

Location: East flank of the Patagonia Mountains, T. 23 S., R. 16 E. (unsurveyed).

Topographic map: Lochiel 15-minute quadrangle.

Geologic map: Wilson, Moore, and O'Haire, 1960, Geologic map of Pima and Santa Cruz Counties, scale 1:375,000.

Access: From Nogales, about 5 miles northeast on State Highway 82 to junction with light-duty road at the Santa Cruz River; from there, 10 miles east across Patagonia Mountains to dirt roads that lead to placers in Mowry Wash.

Extent: Small placers are found in gravels of Mowry Wash and some of its tributaries. Schrader (1915, p. 348) locates four areas where gold was recovered from the gravels: at the east edge of the Quajolote Flat in gravels about 5 feet thick (in Providencia Canyon, near the Four Metals mine); near the head of Mowry Wash in the gravels in a southside tributary (near the old Winifred mine); and in two northside tributaries southeast of Mowry. Placers were also found in Quajolote Wash downstream from the old Mowry smelter; this wash is not named on the topographic map. The gold recovered from the Patagonia placers is generally associated with black sand and in size is generally less than 0.1 inch diameter. One 2-ounce nugget and several smaller nuggets were found.

Production history: The placers in the Patagonia district have a recorded production of about 100 ounces, about equal to that of the Oro Blanco district, although they probably were not so important in the early history of the region.

Source: The detrital gold was apparently freed by erosion of silver, lead, and copper ores that contain very minor amounts of gold. The host rocks for the ore bodies in the drainage area of Mowry Wash are Precambrian quartz monzonite at the Four Metals mine and monzonite and Carboniferous limestone at the Mowry mine.

Literature:

Mining and Scientific Press, 1908: Reports placer-mining activity; production per man per month.

Schrader, 1915: Location; placer-mining activity; average yield per day per man; production.

Wilson, 1961: Quotes Schrader; adds information on placer mining in 1933; size of gold particles.

54. TYNDALL-PALMETTO-HARSHAW DISTRICTS

Location: Low hills on both sides of Sonoita Creek south of the Santa Rita Mountains, Tps. 21 and 22 S., Rs. 14 and 15 E.

Topographic maps: Mount Wrightson and Nogales 15-minute quadrangles.

Geologic maps:

Drewes, 1971a, Geologic map of the Mount Wrightson quadrangle, Santa Cruz and Pima Counties, Arizona, scale 1:48,000.

Wilson, Moore, and O'Haire, 1960, Geologic map of Pima and Santa Cruz Counties, scale 1:375,000.

Extent: Three minor placers are on the west flank of the Patagonia Mountains in the Tyndall, Palmetto, and Harshaw districts. The only information found relates to the location of the placers given by Schrader (1915, p. 220, 279) for the Tyndall and Harshaw districts, and by Wilson (1961, p. 84), for the Palmetto district.

The placers in the Tyndall district are found in the gravels in the open basin at the head of the tributary to Ash Canyon in the SW1/4 of sec. 35 and on both sides of the township line between Tps. 21 and 22 S., R. 14 E. (Mount Wrightson quadrangle).

The placers in the Harshaw district are in Quaternary gravels on a mesa southeast of the junction of Sonoita Creek and Alum Canyon in secs. 13 and 24, T. 22 S., R. 15 E. (Mount Wrightson quadrangle).

The placers in the Palmetto district are in Three R Canyon 2½ miles northwest of the Three R mine in sec. 27, T. 22 S., R. 15 E. (Nogales quadrangle); the placers were worked for 1 month in 1927 and abandoned.

Production history: The Tyndall and Palmetto deposits have had no recorded production during the 20th century, and are relatively unknown. A small production was made from the Palmetto district.

Source: The detrital gold was probably eroded from small gold-bearing veins in the vicinity of the placers.

Literature:

Schrader, 1915: Tyndall district—location. Harshaw district—location; accessory minerals; placer-mining activity.

Wilson, 1961: Tyndall district—quotes Schrader. Harshaw district—quotes Schrader. Palmetto district—location; placer-mining activity in 1927.

YAVAPAI COUNTY

[The most productive placers in Arizona are in the high mountainous region of south-central Yavapai County. Most of the placers are concentrated on the slopes of the Bradshaw Mountains in the vicinity of many small lode deposits. Because of the large number of small mining districts that include parts of gold-bearing streams. I have grouped the placers on the basis of drainage areas rather than formal mining districts]

55. LYNX CREEK DRAINAGE AREA

Location: North flank of the Bradshaw Mountains and the south side of Lonesome Valley, Tps., 13 and 14 N., R. 1 W.; T. 14 N., R. 1 E.

Topographic maps: Prescott and Mount Union 15-minute quadrangles.

Geologic maps:

Anderson and Blacet, 1972b, Geologic map of the Mount Union quadrangle, Yavapai County, Arizona, scale 1:62,500.

Krieger, 1965, Geologic map and sections of the Prescott quadrangle, Arizona, (pl. 1), scale 1:48,000.

Access: From Prescott, State Highway 69 east parallels the lower course of Lynx Creek, and light-duty roads lead south from State Highway 69 at Prescott to many points along the upper reaches of Lynx Creek.

Extent: Placers occur along the entire length of Lynx Creek from near the headwaters at Walker, 7 miles southeast of Prescott, downstream to the junction of Lynx Creek with the Aqua Fria River, 13 miles east of Prescott.

The placers along the upper reaches of Lynx Creek (in the Walker district) occur in the main creek and along its tributaries from near Walker (sec. 34, T. 13 N., R. 1 W., Mount Union quadrangle) downstream for a distance of about 8 miles to the lower dam area (sec. 22, T. 14 N., R. 1 W., Prescott quadrangle). This part of Lynx Creek flows across Precambrian rock, and the gold occurs in thin gravels on narrow benches or bars.

The placers in lower Lynx Creek occur in the east-trending part of the creek from the area around the lower dam, east to the junction with the Aqua Fria River (sec. 34, T. 14 N., R 1 E., just east of the Prescott quadrangle). Gold occurs in the recent alluvium

at the bottom of the steep-walled gulch cut into Tertiary conglomerate. The placer gravels attain a minimum width of more than
one-eighth of a mile and have a thickness of 8–24 feet; a rich pay
streak 4 feet thick was found 2 feet above the conglomerate bedrock.

An area called the Nugget Patch, south of the lower dam on
Lynx Creek (sec. 3, T. 13 N., R. 1 W., Prescott quadrangle), is
said to contain gold in black sands that were probably derived
from quartz veins in the underlying Precambrian gabbro (Krieger,
1965, p. 114).

Production history: Lynx Creek is the most productive gold-bearing
stream in Arizona, although other districts (La Paz, Yuma County;
Weaver, Yavapai County) have yielded more gold from alluvial
fans, flats, and arroyos. The Lynx Creek placers were discovered
in May 1863 by Sam Miller and four other prospectors of the group
led by Captain Joe Walker. Sam Miller reportedly panned $4.80
in gold from a gravel bank along Lynx Creek; on May 10, 1863,
the party organized the first mining district in Yavapai County, which
they called the "Pioneer District." The Walker quartz mining
district was formed November 24, 1863. Production from the Lynx
Creek placers before 1900 is generally estimated at about $1 million,
although some writers estimate $2 million.

During the 20th century the placers in the lower section of Lynx
Creek have been the most actively mined. Large-scale placer mining
was done by dredges operating along 5 miles of lower Lynx Creek
from the lower dam in sec. 22, T. 14 N., R. 1 W., to the vicinity
of Fain's Ranch in sec. 28, T. 14 N., R. 1 E. (Prescott quadrangle).
The Calari Dredging Co. worked placer ground in 1933 below the
lower dam that averaged 32 cents per cubic yard. In late 1939 the
Rock Castle Placer Mines Co. used a dryland dredge to work the
bench gravels in this area. From 1934 to 1940 (in particular the
years 1938–39) the Lynx Creek Placer Mining Co. worked the
gravels on the Fitzmaurice property, which extends from secs. 22–24,
T. 14 N., R. 1 W., through sec. 19, T. 14 N., R. 1 E.; this dredge
was the largest single producer in Arizona.

Most of the placer mining in the area of upper Lynx Creek was
small-scale rocking and sluicing, but a few larger scale placer operations were attempted, especially in that part of upper Lynx Creek
just downstream from the old Highway Bridge (NW ¼ sec. 33, T. 14
N., R. 1 W.). During the period 1940–41 gravels were worked in the
area called Bigelow Flat to about half a mile below the bridge, a
distance of about 3 miles.

Source: The placer gold in Lynx Creek was derived from numerous
widely scattered small gold-quartz veins in adjacent parts of the

Bradshaw Mountains. Mineralization in the Bradshaw Mountains is both Precambrian and Laramide in age, and placers have been derived from veins of both ages. In the Walker area, the gold-quartz veins are associated with a small stock of granodiorite that recent work has shown to be of Laramide age (64 m.y.; Anderson, 1968, p. 1169). Most of the gold in Lynx Creek is thought to have been derived from the gold veins in the Walker area. The gold found along the creek varies from coarse nuggets to 4 ounces in the upper reaches of the creek to fine gold along the lower reaches of Lynx Creek. The gold-silver ratio in the nuggets increases downstream.

Literature:

Allen, 1922: Location; production; placer-mining operations during the periods 1907–9 and 1918–19.

Blake, 1899: Location; placer-mining operations; problems; gold values per cubic yard.

Burchard, 1882: Production estimate (1863–81).

———1884: Brief history of early placer mining; locates placer ground near lode mines on upper Lynx Creek.

———1885: Production estimates and production for 1884.

De Wolf, 1916: Reports four hydraulic giants installed at Lynx Creek.

Gardner and Johnson, 1935: Depth of gravel; placer-mining operations on upper Lynx Creek.

Gardner and Allsman, 1938: Lists—placer-mining techniques; depth and characteristics of gravel mined; depth of bedrock mined; percent of gold recovery.

Koschmann and Bergendahl, 1968: History; placer-mining operations; production.

Krieger, 1965: History; location of placer-mining operations; bedrock geology.

Lindgren, 1926: History; production; extent of placers, character and value of gold; source.

Raymond, 1872: Walker district—extent of placers; placer-mining problems on upper Lynx Creek.

Wilson, 1961: Location; history; production; dredging operations to 1949. Small-scale operations in 1933; geology of gold in gravels.

Wood, 1929: History of placer discovery.

56. HASSAYAMPA RIVER DRAINAGE AREA

Location: West flank of the Bradshaw Mountains, Tps. 8–13 N., Rs. 2–5 W.

Topographic maps: Congress 30-minute quadrangle; Wagoner NE 15-minute quadrangle (covers only NE ¼ of quadrangle); Kirkland

E. 15-minute quadrangle (covers only E ½ of quadrangle); Mount Union 15-minute quadrangle.

Geologic maps:

Arizona Bureau of Mines, 1958, Geologic map of Yavapai County, scale 1:375,000.

Anderson and Blacet, 1972b, Geologic map of the Mount Union quadrangle, scale 1:62,500.

Access: U.S. Highway 89 leads south to Congress from Prescott west of the placer areas along Hassayampa Creek. Dirt roads lead east from U.S. Highway 89 to the different placer areas along the Hassayampa.

Extent: Placers are found along most of the Hassayampa River and in many tributaries from Groom Creek near the headwaters, downstream to Blue Tank Wash, a tributary near Wickenburg.

Upper Hassayampa River (Hassayampa district): In the headwaters of the Hassayampa River, placers are found along Groom Creek, the Hassayampa River, and small side gulches (T. 13 N., R. 2 W., Mount Union quadrangle). I have found no description of the gold-bearing gravels in Groom Creek and nearby parts of the Hassayampa River.

Central Hassayampa River (Wagoner, Walnut Grove, and Tiger districts): The central part of the Hassayampa drainage area, near Walnut Grove and Wagoner, includes the Hassayampa River, tributaries on the west side (Placerita and French Gulches), tributaries on the east side (Blind Indian, Milk, Minnehaha, Cherry and Oak Creeks). The gravels in the river near Walnut Grove contain many boulders but no clay; the gold is described as flake gold. The Hassayampa River was most actively worked between 1885 and 1890; during that time a dam was built (near the junction of Cherry Creek with the Hassayampa—sec. 23, T. 10 N., R. 3 W., Wagoner quadrangle) to permit hydraulic mining in Rich Hill (10 miles west) and large-scale operations on the Hassayampa downstream from Wagoner. The dam failed in 1890, killing 150 people and flooding the downstream section of the Hassayampa.

Gravels were mined along the upper parts and side gulches of Placerita Gulch (approximately sec. 14, T. 11 N., R. 4 W., Congress quadrangle), near the junction of Placerita and French Gulches (secs. 7 and 18, T. 11 N., R. 3 W., Kirkland quadrangle), and on French Gulch, 1 mile southeast of Zonia (sec. 17 or 18, T. 11 N., R. 3 W.). Much of the gold in these gulches is fairly coarse and many ¼- and ½-ounce nuggets were recovered.

Placers have also been found in Blind Indian and Milk Creeks on the east side of the Hassayampa (Tps. 10 and 11 N., Rs. 2 and 3

W.), but these were not worked so extensively as the placers in Placerita and French Gulches. Placer gold was reportedly found on Slate and Milk Creeks in beds of volcanic agglomerate that were hydraulicked before 1905.

South of Blind Indian and Milk Creeks, small placers were worked in Minnehaha, Cherry, and Oak Creeks. Placers were found near the headwaters of Minnehaha Creek in Minehaha Flat (unsurveyed secs. 19, 30, 31, T. 10 N., R. 1 W., Crown King quadrangle) and on Oak Creek, 1 mile below Fentons Ranch (sec. 3 or 9, T. 9 N., R. 2 W., Crown King quadrangle).

Lower Hassayampa River (Black Rock and Blue Tanks districts): Small placers are found in the Black Rock region in T. 8 N., R. 3 W. (Congress quadrangle) and on the Hassayampa River near the mouth of Blue Tank Wash in T. 7 N., R. 5 W., near the Maricopa County-Yavapai County boundary.

Production history: Upper Hassayampa River (Hassayampa district): The placers in Groom Creek were discovered in the 1860's and actively worked in the 1880's. Sparks (1917) estimated $3 million production in placer gold from Groom Creek, but this estimate is probably grossly high. During the 1930's this northern region was placered on a small scale by many individuals, and from 1939 to 1942, a dragline dredge on the Hobbs property (unlocated) on the Hassayampa River recovered several hundred ounces of placer gold.

Central Hassayampa River (Wagoner, Walnut Grove, and Tiger districts): Most placer mining in the central region was done by individuals using drywash machines.

Lower Hassayampa River (Black Rock and Blue Tanks districts): Minor amounts of placer gold were recovered intermittently from this region and from the Hassayampa River in Maricopa County.

Source: The tributaries of the Hassayampa River drain a wide area of mineralized terrain. The ore deposits that contributed the gold found along the Hassayampa and its tributaries are of both Precambrian and Tertiary age, and it is difficult to demonstrate which vein or vein systems provided the source of the placer gold. Lindgren (1926) summarizes the physical characteristics and probable age of many of the veins in the area.

Literature:

Allen, 1922: Location; names placer-bearing tributary creeks.

Blake, 1899: Notes presence of placer gold.

Browne, 1868: Describes discovery of placers on the Hassayampa River.

Burchard, 1882: General history of placer mining along the Hassayampa.

————1883: History and placer-mining activity at Placeritas.

Church, 1887: Reports progress on Walnut Grove Dam.

De Wolf, 1916: Reports plans to rebuild Walnut Grove Dam.

Engineering and Mining Journal, 1933c: Results of sampling placer ground at Minnehaha placers.

————1890: Reports failure of Walnut Grove Dam.

Girand, 1932: Describes drywash machine used at Walnut Grove; characteristics of gravels; size of gold recovered.

Jagger and Palache, 1905: Reports gold in Blind Indian and Milk Creek.

Koschmann and Bergendahl, 1968: Placer-gold production.

Lindgren, 1926: Notes placers in Minnehaha Flat; locates placers on upper Hassayampa.

Raymond, 1872: Location; extent; placer-mining techniques.

Sparks, 1917: Production estimates for Groom Creek.

U.S. Bureau of Mines, 1926–31: Names claims and creeks where placer mining is active.

————1934: Names claims and creeks where placer mining was active.

Wilson, 1961: Hassayampa placers—location; production history; placer-mining activity during the period 1932–33; source. Groom Creek—location; production estimates; source. Placerita placers—early history; placer-mining activity during the period 1932–33; size of nuggets.

57. BIG BUG CREEK DRAINAGE AREA

Location: East flank of the Bradshaw Mountains, Tps. 12 and 13 N., R. 1 W.; Tps. 12 and 13 N., Rs. 1 and 2 E.

Topographic maps: Mount Union and Mayer 15-minute quadrangles.

Geologic maps:

Anderson and Blacet, 1972a, Geologic map of the Mayer quadrangle, Yavapai County, Arizona, scale 1:62,500.

————1972b, Geologic map of the Mount Union quadrangle, Yavapai County, Arizona, scale 1:62,500.

Access: State Highway 69 leads east and south from Prescott to Humboldt, Poland Junction, and Mayer. Placers are found adjacent to the highway near Mayer and are easily accessible by dirt roads from the highway.

Extent: Placers are found in stream gravels and gravel-covered mesas in a roughly triangular area that extends for about 20 miles east and northeast from the head of Big Bug Creek. Most placer-mining activity was concentrated in the part of Big Bug Creek, tributary gulches and gravel benches in the area bounded by McCabe, Humboldt, and Mayer (Tps. 12 and 13 N., R. 1 E., Mayer and Mount Union quadrangles). Apparently, there was little or no mining in the part of Big Bug Creek downstream from Mayer. Small placers were worked in the upper reaches of Big Bug Creek below the Mesa mine, about 2 miles south-southeast of Walker, on the south side of the creek (approximately sec. 9, T. 12 N., R 1 W., unsurveyed, Mount Union quadrangle); in Eugene Gulch, a major tributary to Big Bug Creek (south edge of T. 13 N., R. 1 W., Mount Union quadrangle), and in Chaparral Gulch and other small gulches near McCabe (W½ of T. 13 N., R. 1 E., Mount Union quadrangle).

Production history: The placers in the drainage area of Big Bug Creek were discovered during the 1860's, but the greatest activity in placer mining was during the 1880's (Wilson, 1961, p. 48). Wilson states that no estimate of early production is available, but recorded production for the 20th century (second only to Lynx Creek in Yavapai County) indicates early production probably was large.

In the eastern part of the area, north and northwest of Mayer, many small-scale and some large-scale placer operations have been active during the 20th century. The Shank and Savoy property (or the Savoy property) was active for many years; this placer is in a side gulch on the west side of Big Bug Creek and extends about 3 miles northwest of Mayer (probably Grapevine Gulch, sec. 8, T. 12 N., R. 1 E., Mount Union quadrangle). The gold-bearing gravels overlying cemented gravels were bouldery to sandy, with little clay. The gold here was irregularly distributed, flat to round particles that were as much as 50 cents in value. This placer was worked by a number of companies and probably produced most of the placer gold reported from the district during the 1930's and 1940's.

Other placer-mining activity was concentrated in the area surrounding Mayer (especially sec. 22, T. 12 N., R. 1 E., Mayer quadrangle) where the placers are found in a wide gravel-covered area between outcrops of metamorphosed Precambrian volcanic rocks. The area was worked intensively for many years and during the 1932–33 season was reworked by individuals who mined the gravels by tunneling and packing the pay dirt to sluices, rockers, or small concentrating machines.

Source: The placers in the Big Bug district originated by erosion of many small, and some large, gold-bearing veins on the east flank of the Bradshaw Mountains. Some veins, such as the Mesa mine in the upper part of Big Bug Creek, and several veins in the vicinity of McCabe, are considered to be Precambrian in age; other veins, in the vicinity of Poland and Providence, are considered to be later, probably Laramide (Lindgren, 1926, p. 127).

Literature:

Allen, 1922: Placer operations during the period 1900–1901; reason for failure.

Arizona Mining Journal, 1919: Placer-mining activity in Eugenia (Eugene) Gulch.

Koschmann and Bergendahl, 1968: Placer gold production.

Lindgren, 1926: Locates small placers on upper Big Bug Creek.

Randolph, 1903: Notes presence of placer gold.

Wilson, 1961: Location; extent and character of gold-bearing gravel; placer-mining operations.

58. TURKEY CREEK DRAINAGE AREA

Location: East flank of the Bradshaw Mountains, T. 11 N., R. 1 E.

Topographic maps: Mount Union and Mayer 15-minute quadrangles.

Geologic maps:

Anderson and Blacet, 1972a, Geologic map of the Mayer quadrangle, Yavapai County, Arizona, scale 1:62:500.

——1972b, Geologic map of the Mount Union quadrangle, Yavapai County, Arizona, scale 1:62,500.

Access: About 2 miles south of Mayer, a light-duty road leads south from State Highway 69 about 8 miles to the Turkey Creek area; about 11 miles south of Mayer, the road to Cleator leads west about 3 miles from State Highway 69 and a dirt road leads north to Turkey Creek area.

Extent: Small placer deposits near Turkey Creek Station (once located 1 mile north of Cleator, sec. 35, T. 11 N., R. 1 E., unsurveyed, Mayer quadrangle) were drywashed for many years; placers are said to occur in the vicinity of Pine Flat, along the upper part of Turkey Creek near the Cunningham mine (approximately sec. 5, T. 11 N., R. 1 W., unsurveyed, Mount Union quadrangle). Two small placers are located on Turkey Creek, midway between the two placers mentioned above, which are about 1.6 and 2.9 miles upstream from Turkey Creek Station (unsurveyed area, SE¼ Mount Union quadrangle). Placers were also worked in terrace gravels along Turkey Creek downstream from the Golden Turkey mine (approximately sec. 12, T. 10 N., R. 1 E., unsurveyed).

Production history: The placers in Turkey Creek were worked intermittently between 1908 and 1941, and small amounts of gold recovered. The placer gold credited to the Peck district by the U.S. Bureau of Mines probably comes from this area.

Source: The placers in the Turkey Creek drainage probably were derived from gold veins in close proximity to the various small placers. Lindgren (1926, p. 156) states that Precambrian gold veins at Turkey

Creek Station are the source of the gold in the shallow placers on the flat below the veins.

Literature:

Blacet, 1968: Map locates placers on Turkey Creek.

Guiteras, 1936: Placer-mining operations in 1936.

Lindgren, 1926: Locates placers on Turkey Creek.

59. BLACK CANYON DRAINAGE AREA

Location: East flank of the Bradshaw Mountains, Tps. 9½ to 10 N., Rs. 1 and 2 E.

Topographic maps: Mayer and Bumble Bee 15-minute quadrangles.

Geologic maps:

Anderson and Blacet, 1972a, Geologic map of the Mayer quadrangle, Yavapai County, Arizona, scale 1:62,500.

Arizona Bureau of Mines, 1958, Geologic map of Yavapai County, scale 1:375,000.

Access: State Highway 69 parallels Black Canyon between Bumble Bee and Arrastre Creek.

Extent: Placers occur along the Black Canyon segment of Turkey Creek between Arrastre Creek and Poland Creek and have been worked upstream to the vicinity of Cleator. Placers were also mined in American and Mexican Gulches where Bumble Bee Creek enters Black Canyon.

Placers were worked in Black Canyon below Howard's Copper mine (sec. 31, T. 10 N., R. 2 E., Bumble Bee quadrangle), where before 1922, one man reportedly produced about $20,000 in gold, probably at a spot about 1 mile downstream from the mine. Gold was also recovered from a gravel bar in Black Canyon about 3 miles south of Bumble Bee (probably sec. 32, T. 9½ N., R. 2 E., Bumble Bee quadrangle).

Production history: The placers in the Black Canyon area have been worked intermittently since the latter part of the 1800's. The amount of gold recovered from the Black Canyon placers was not large compared with other placers in Yavapai County but was appreciably more than that recovered along the upper part of Turkey Creek.

Source: The placers in the Black Canyon drainage were derived from gold veins that are apparently of both Precambrian and Late Cretaceous to early Tertiary (Laramide) age. Precambrian gold veins occur in the vicinity of Bumble Bee (on Bland Hill and at the Gillespie mine), and "Laramide" veins occur throughout the district, in particular, at the Thunderbolt mine in Black Canyon.

Literature:

Browne, 1868: Reports platinum in placers.

Burchard, 1882: Reports placer occurrence.

Lindgren, 1926: Locates placers.

Wilson, 1961: Location; extent of placer-mining activity during the period 1932–33.

60. HUMBUG CREEK DRAINAGE AREA

Location: South flank of the Bradshaw Mountains, Tps. 7–10 N., R. 1 E.; Rs. 1 and 2 W.

Topographic maps: Bradshaw Mountains 30-minute quadrangle; Prescott 2-degree sheet, Army Map Service; Crown King 15-minute quadrangle (covers only N½ of quadrangle); Governors Peak 7½-minute quadrangle.

Geologic maps:

Arizona Bureau of Mines, 1958, Geologic map of Yavapai County, scale 1:375,000.

Lindgren, 1926, Geologic map of the Bradshaw Mountains quadrangle, Arizona (pl. 2), scale 1:125,000.

Access: Light-duty roads lead northwest to Humbug Creek from State Highway 69 north of Lake Pleasant.

Extent: Placers are found in many of the creeks that drain the south flank of the Bradshaw Mountains. Humbug, French, and Cow Creeks (Tps. 7–9 N., R. 1 W., R. 1 E.) reportedly contain placer gold throughout 20 miles of the drainage area. Buckhorn Gulch, Castle Creek, and small tributaries have been placered along the upper reaches in the vicinity of Copperpolis (T. 8 N., R. 2 W.).

Production history: The production from the placers found in the Humbug Creek drainage area is generally listed by the U.S. Bureau of Mines under Humbug, Tip Top, Tiger, Silver Mountain, Castle Creek, and White Picacho districts. Most of the placer production has been small and intermittent. During the 1800's, placers in Rockwall, Carpenter, and Swilling Gulches, tributaries to Humbug Creek (T. 8 N., R. 1 W.–R. 1 E., Bradshaw Mountains quadrangle) were reportedly very productive but were exhausted early in the 1900's. During the 1890's, the Humbug Hydraulic Mining Works planned large-scale placer mining on lower Humbug Creek and constructed a dam in sec. 6, T. 7 N., R 1 E., (Governors Peak quadrangle). This enterprise met with failure because the size of the gold was too small for the recovery methods used. The Star Placer on Humbug Creek (sec. 13, T. 8 N., R. 1 W.) and the Horseshoe placer (unlocated) on French Creek were worked after 1900.

Small amounts of placer gold were recovered from Buckhorn Gulch, Castle Creek, American Gulch, and Todos Santos Creek (unlocated, White Picacho district) . The John D. placer, on Castle Creek (sec. 9, T. 8 N., R. 2 W.), has been worked in the 20th century. Plans were made to dredge parts of Buckhorn Gulch, a tributary to Castle Creek, but apparently the operation did not materialize.

Source: Gold placers along the upper reaches of Humbug Creek and especially in Carpenter, Swilling, and Rockwall gulches, were probably derived from gold veins that intersect and offset rhyolite porphyry dikes (described by Lindgren, 1926, p. 179); association with these dikes is considered indicative of a "Laramide" age for gold mineralization in the Bradshaw Mountains.

Gold placers along Castle Creek and tributaries were derived from Precambrian ore deposits that predominate in that area.

Literature:

Allen, 1922: Placer operations in the 1890's; reasons for failure.

Burchard, 1885: Reports placer mining on Coso Creek and tributaries (unlocated).

De Wolf, 1916: Reports activity at Humbug Creek.

Lindgren, 1926: Notes production from tributaries to Humbug Creek.

Wilson, 1961: Extent and thickness of gold-bearing gravels; distribution of gold in gravels; placer-mining activity during the period 1932–33.

61. WEAVER (RICH HILL) DISTRICT

Location: South flank of the Weaver Mountains. Tps. 9 and 10 N., Rs. 4 and 5 W.

Topographic maps: Congress 30-minute quadrangle; Prescott 2-degree sheet, Army Map Service.

Geologic map: Arizona Bureau of Mines, 1958, Geologic map of Yavapai County, scale 1:375,000.

Access: From Prescott, 38 miles southwest on U.S. Highway 89 to light-duty road leading 8 miles east to Rich Hill. Dirt roads lead to placer ground.

Extent: The Weaver placer area covers about 40 square miles on the south flank of the Weaver Mountains. The most important placer area in production and placer-mining activity is the area at the top of Rich Hill, parts of the sides of the hill, and gravels along Weaver and Antelope Creeks. This district is just north of Octave and east of Stanton (at the intersection of T. 10 N., R. 5 W.; Tps. 9 and 10 N., R. 4 W.). At the top of Rich Hill, gold was found under boulders and in crevices in the granite bedrock, where it was quickly gathered by prospectors during the early years after the discovery of the placers. Below Rich Hill, in Antelope and Weaver Creeks, the gold was found in reconcentrated stream gravels, a few feet thick to more than 50 feet thick, that contained numerous large bolders.

Other placers are found west of this area, in the vicinity of the

Planet and Saturn mines (sec. 21, T. 10 N., R. 5 W.). This area probably produced the placer gold attributed to the Martinez district.

Production history: The Rich Hill placers were discovered by a party of prospectors led by Captain Pauline Weaver in 1863 or 1864 (one account reports 1862 as the year of discovery) about the same time as the discovery of the Lynx Creek placers. According to many reports, a Mexican in the party found loose gold on the top of Rich Hill while looking for a stray animal. Immense excitement and intense mining activity followed the discovery. Within 3 months, $108,000 in gold ranging in size from a pinhead to large nuggets worth hundreds of dollars was recovered, and within 5 years, $500,000 in placer gold was recovered. By 1883, about $1 million in placer gold was recovered. The placers have been worked extensively since the discovery, but because of the nature of the gravels, few large-scale operations have been attempted. Most of the mining has been done by drywashers, pans, rockers, and sluices, although some miners used power shovels and dry-separation plants.

Source: There has been no detailed geologic study of the Weaver Mountains, therefore details of the nature of gold-bearing veins are not known. The mountains are composed principally of Precambrian granites and schists that contain numerous gold-bearing veins considered to be of Laramide age. Some of these veins in the vicinity of the placers have been mined for their gold content, and it is probable that the placers were probably derived from these and other similar veins in the vicinity.

Literature:

Allen, 1922: Location; history; size of gold nuggets; production; distribution of different size particles of gold in different types of placer gravels.

Blake, 1899: Location; size of nuggets; production.

Blandy, 1897: Production estimates.

Burchard, 1882: History; production estimates (1863–81).

———1885: History; early placer-mining activity.

Esenwein, 1958: Describes new placer discovery; accessory minerals in placer.

Gardner and Johnson, 1935: Placer-mining operations at Octave in 1931.

Hamilton, 1884: Production estimates; history.

Heikes and Yale, 1913: Area and thickness of placer gravels; size of gold nugget; distribution of gold in gravels; production from 1905 to 1912.

Heineman, 1931: Physical characteristics of nugget from Red Bank placer.

Hodge, 1877: Production from Rich Hill during the first 3 months after discovery.

Koschmann and Bergendahl, 1968: History; production.

Mining Journal, 1938c: Details of placer-mining apparatus at Thunderbird property.

———1938e: Describes gravels at Thunderbird placer.

Sawyer, 1932: Detailed description of sampling techniques; includes description of placer gravels.

Watson, 1918: Discusses general history of placer mining at Rich Hill.

Wilson, 1961: Describes placers in Yavapai County (p. 38–57). Location; extent; character of gold; distribution of gold-bearing gravels; placer-mining history and activity.

62. MODEL AND KIRKLAND PLACERS

Location: East flank of the Weaver Mountains in Peeples Valley, Tps. 11 and 12 N., Rs. 4 and 5 W.

Topographic maps: Congress 30-minute quadrangle; Prescott 2-degree sheet, Army Map Service.

Geologic map: Arizona Bureau of Mines, 1958, Geologic map of Yavapai County, scale 1:375,000.

Access: From Prescott, 26 miles southwest on U.S. Highway 89 to Peeples Valley. Dirt roads lead from the highway east to Kirkland Creek and west to Model.

Extent: Small placers are found on the west side of Peeples Valley in the vicinity of Model Creek and on the east side of Peeples Valley in Kirkland Creek. The placers in Model Creek and other gulches occur in small local basins or channels for about three-quarters of a mile on each side of Model Creek on the pediment area between the Weaver Mountains and Peeples Valley. This area cannot be accurately located because there are no large-scale maps, but it is probably near the Model mine in T. 11 N., R. 5 W. (Congress quadrangle). The Columbia placer is near the old Monitor mine in sec. 21, T. 11 N., R. 5 W.

The location of the placers in Kirkland Creek is not certain, because U.S. Bureau of Mines records indicate only that the placers are on Kirkland Creek near Kirkland (T. 12 N., R. 4 W.). The headwaters of Kirkland Creek drain the northwest flank of the Weaver Mountains, opposite the Placeritas and French Creek placers, and the Kirkland placers probably are between the mountain slope and Kirkland.

Production history: Very little is known about the production of the placers in Model Creek; the area was active in 1933, but production for that year was included with production from the Weaver district by the U.S. Bureau of Mines.

The placers in Kirkland Creek were worked on a small scale for a few years during the period 1934–57.

Source: Unknown. Presumably, the gold was derived from small gold-bearing veins in the immediate vicinity of the placers, but, because the area has not been studied in detail, no information is known about the occurrence of the gold-bearing veins.

Literature:

Wilson, 1961: Model placers: location; extent and character of gold-bearing gravels; size of gold particles; placer-mining activity during the period 1932–33.

63. COPPER BASIN DISTRICT

Location: Northeast and southwest flanks of the Sierra Prieta, Tps. 13 and 14 N., Rs. 3 and 4 W.

Topographic maps: Kirkland 15-minute quadrangle (covers only E½ of quadrangle); Iron Springs 15-minute quadrangle; Congress 30-minute quadrangle; Prescott 2-degree sheet, Army Map Service.

Geologic maps:

Krieger, 1967, Reconnaissance geologic map of the Iron Springs quadrangle, Yavapai County. Arizona, scale 1:62,500.

Arizona Bureau of Mines, 1958, Geologic map of Yavapai County, scale 1:375,000.

Access: From Prescott, about 12 miles southwest on light-duty road paralleling Aspen Creek to Copper Basin; a light-duty road leads north across the crest of the Sierra Prieta to Miller Creek and Thumb Butte.

Extent: Placers in the Copper Basin district are found in gulches that dissect the pediment slope on the southwest flank of the Sierra Prieta and in the gravel-floored plain between this pediment and Skull Valley. A small amount of placer gold was recovered between 1933 and 1940 near Thumb Butte on the northeast slope of the mountains.

In the immediate vicinity of Copper Basin (secs. 20 and 21, T. 13 N., R. 3 W., Kirkland quadrangle) placers have been mined from terrace gravels and stream gravels in and near Copper Basin Wash and tributary gulches; tht terrace gravels contain oxidized copper deposits and some gold placers. These deposits contain gold fragments that are wiry to angular and noticeably coarser than the placers away from the mountains. Oxidized copper minerals are common in placers in the upper part of the Copper Basin Wash.

Other placers near the mountains include deposits in Mexican Gulch, Copper Creek, and Spruce Canyon. Mexican Gulch is not located on the topographic maps, but Wilson (1961, p. 47) states that it is 2½ miles from Skull Valley and probably near the low hills at the east edge of T. 14 N., R. 4 W. The deposit in Mexican Gulch had an over burden of 10–15 feet of soil and sand on top of pay gravel 2 feet thick that contained appreciable amounts of clay. The gold recovered from this gulch was commonly the size of mustard seed to $3 nuggets, but some $15 nuggets were found (valued at $20.67 per oz). Spruce Canyon is on the northwest slope of the Sierra Prieta, about 5 miles northeast of Skull Valley (T. 14 N., R. 4 W., Iron Springs quadrangle); results of sampling in 1933 indicated values of 56 cents to $2.12 per yard. Copper Creek, a tributary to the Hassayampa River, is southeast of Copper Basin (Tps. 12½ and 13 N., R. 3 W., Kirkland quadrangle), but the exact location of the placers is unknown.

Apparently, the richest placers are on the gravel-floored plain west and south of Copper Basin (T. 13 N., R. 4 W.); most of this area is shown only on the Congress 30-minute quadrangle, which does not show the details necessary for accurate location of the placers. The gravels in this area are characterized by small boulders and more clay than the gravels near the mountains. The gold in these gravels occurs as particles worth less than 25 cents; larger nuggets are rare. The placers contain small particles of cinnabar and natural amalgam probably derived from cinnabar veins in the Copper Basin area.

Production history: The placers in the Copper Basin district were worked intermittently until 1931. At that time a few companies started placer mining on a large scale in various localities in Copper Basin. The Aztec placer claim (and other unlocated claims) was active in 1931. The Aztec deposit in Copper Basin Wash (sec. 21, T. 13 N., R. 3 W.) is in terrace gravels cemented by copper carbonate and oxide minerals. Most of the large-scale placer mining was apparently concentrated in the southwestern part of the placer field; during the early 1930's the Forbach and Easton Co., the Skull Valley Corp., and the Gold Star Placer Co. worked these deposits. Unfortunately, lack of detailed maps precludes location of any of these claims.

Source: The source of the placer gold in Copper Basin is reported to be gold-bearing tourmaline-quartz veins of Precambrian age (Johnston and Lowell, 1961).

Literature:

Engineering and Mining Journal, 1933a: Results of sampling placer ground in Spruce Canyon.

Gardner and Johnson, 1935: Depth of gravel; type of bedrock; accessory minerals in placer gravels; size of nuggets; placer operations.

Girand, 1932: Describes gravels in Mexican Gulch.

Johnston and Lowell, 1961: General bedrock geology; source of placer gold.

Mining Journal, 1932a: Reports recovery of nugget weighing 6 ounces from Chase placer.

U.S. Bureau of Mines, 1929, 1931: Names placer-bearing creeks and placer claims.

————1934–35: Names placer creeks; placer-mining operations.

Wilson, 1961: Location; extent and character of gold-bearing gravels; size of gold particles; accessory minerals in placers; placer-mining activity during the period 1932–33.

64. GRANITE CREEK

Location: North flank of the Bradshaw Mountains and in Chino Valley. Tps. 13–17 N., R. 2 W.

Topographic maps: All 15-minute quadrangles—Mount Union, Prescott, Paulden.

Geologic map: Krieger, 1965, Geologic map and sections of the Prescott and Paulden quadrangles, Arizona (pls. 1, 2), scale 1:48,000.

Access: U.S. Highway 89 parallels Granite Creek in the Bradshaw Mountains, leads north, parallel to Granite Creek 22 miles to Del Rio.

Extent: Granite Creek heads on the north flank of the Bradshaw Mountains and flows north into Chino Valley. Most of the placers are found in the upper drainage of Granite Creek between Prescott and the head of the creek (T. 13 N., R. 2 W., Mount Union and Prescott quadrangles). Gold was recovered from the gravels of the lower reaches of Granite Creek as far north as Granite Dells (sec. 12 T. 14 N., R. 2 W., Prescott quadrangle), and a little gold was recovered from creek gravels near Del Rio (secs. 22 and 23, T. 17 N., R. 2 W., Paulden quadrangle). The gold reported between Granite Dells and Del Rio was found in washes and in pediment and terrace gravels.

Production history: The placers in Granite Creek south of Prescott were discovered in the 1860's and actively worked during the 1880's. Sparks (1917) states that the town of Prescott "owes its present beautiful situation * * * to the fortunes made by placer miners in Granite Creek." This statement is perhaps an exaggeration, but the town is constructed upon some minor placers, as Wilson (1961, p. 56) reports that small nuggets were found in gravels uncovered by building excavations in Prescott. Apparently, the placers on the mountain slope were very rich; one placer miner reportedly recovered $20,000 in placer gold from New England Gulch (probably before 1872), a branch of Granite Creek about 4 miles south of Prescott. The placers are located in that part of the Gulch in the NW¼ sec. 21, T. 13 N., R. 2 W., (Mount Union quadrangle).

Source: According to Lindgren (1926, p. 108), the lodes in the northern foothills of the Bradshaw Mountains are of Precambrian age. The origin of placers located north of Prescott on Granite Creek is unknown.

Literature:

Krieger, 1965: Location of gold-bearing gravels.

Lindgren, 1926: States that gold was recovered at Del Rio.

Sparks, 1917: Notes placers in Granite Creek at Prescott.

Wilson, 1933: Production from New England Gulch.

———1961: Location; history (4th ed., 1933).

65. EUREKA (BAGDAD) DISTRICT

Location: Burro Creek, Santa Maria River, and the vicinity of Bagdad, Tps. 12–15 N., Rs. 8–10 W.

Topographic maps: Prescott 2-degree sheet, Army Map Service; Bagdad 15-minute quadrangle; Congress 30-minute quadrangle.

Geologic maps:

Anderson, Scholz, and Strobell, 1955, Geologic map of the Bagdad area, Yavapai County, Arizona (pl. 3), scale 1:20,000.

Arizona Bureau of Mines, 1958, Geologic map of Yavapai County, scale 1:375,000.

Access: Bagdad is accessible by an improved highway that leads northwest from Hillside, about 35 miles to Bagdad. Access to other areas is unknown.

Extent: Small placers are found in widely scattered areas in the Eureka district. Gold was recovered during the late 1850's at Old Placers near the Cowboy mine, probably located in or near sec. 14, T. 14 N., R. 9 W. (Bagdad quadrangle). Other placers, whose exact location is unknown, are along Burro Creek northwest of Bagdad (Prescott 2-degree sheet) in the vicinity of the Santa Maria River southwest of Bagdad (Prescott 2-degree sheet) and in unnamed gulches northwest of Hillside.

Production history: The placers in the Eureka district have been worked intermittently on a small scale since the late 1850's. In 1875, placers were discovered in the Santa Maria area. These deposits were drywashed that year and yielded a nugget valued (at that time) $50 to

$60. During this century, the deposits throughout this area have been drywashed intermittently with small gold recovery.

Source: The placers near the Cowboy mine were probably derived from erosion of the gold-bearing vein mined there; the other deposits were probably derived from small veins in the vicinity of the placers.

Literature:

Raymond, 1877: Reports placer discovery; describes size of gold in Santa Maria River.

U.S. Bureau of Mines, 1934–35: Placers recovered from Burro Creek, Santa Maria River, and gulches northwest of Hillside.

U.S. Geological Survey, 1922: Placers recovered from Burro Creek and other unnamed creeks.

Wilson, 1961: History; production.

OTHER DISTRICTS

66. LINCOLN CREEK

Placer gold was recovered from Lincoln Creek, apparently in the vicinity of Jerome, in 1932. I cannot locate this creek.

67. POCKET CREEK

Placer gold was recovered in 1942 and 1943 by a single operation. I cannot locate this creek.

68. SQUAW CREEK

Placer gold was recovered in 1933 from a deposit along Squaw Creek, a tributary to the Aqua Fria River (T. 9 N., R. 2 E.) about 2 miles north of the junction of Black Canyon and the Aqua Fria River.

YUMA COUNTY

69. GILA CITY (DOME) DISTRICT

Location: North end of the Gila Mountains; south bank of the Gila River, T. 8 S., R. 21 W.

Topographic maps: Laguna Dam and Dome 7½-minute quadrangles; Laguna 15-minute quadrangle.

Geologic map: Wilson, 1960, Geologic Map of Yuma County, Arizona, scale 1:375,000.

Access: From Yuma, about 13 miles east on State Highway 95 to Blaisdell; light-duty road parallels the Southern Pacific 7 miles east to Dome.

Extent: The Gila City placers occur on the narrow gravel-mantled pediment at the north end of the Gila Mountains formed on a bedrock of Tertiary sedimentary rocks that are faulted against the schist of the main mountain mass. Gold has been found in gulch and bench gravels of Quaternary age that mantle the Tertiary sediments to depths of 15 feet. The area of gold-bearing gravel extends from 1/4 mile east of Dome to 3 miles west of Dome, but most placer mining is centered around Monitor Gulch, 1½ miles west of Dome (sec. 11, T. 8 S., R. 21 W., Laguna Dam quadrangle). Most of the gold in the gravels was found at or near bedrock in gulches, but much gold was recovered from bench gravels in the area. Gravels more than 15 feet above bedrock have not been profitable.

Production history: The Gila City placers were discovered in September 1858 by Colonel Jacob Snively and were actively worked by hundreds to thousands of men until about 1865, when the richest gravels were depleted. Gila City, a placer boom town that lived only 4 years, was near the mouth of Monitor Gulch adjacent to the Southern Pacific (NW¼ sec. 11, T. 8 S., R. 21 W.). Lieutenant Sylvester Mowry, a noted Arizona miner and pioneer, visited the placers in November 1858 and reported that men were recovering $30 to $215 per day; he witnessed $20 in gold washed from eight shovelfuls of dirt by an unexperienced placer miner.

After the initial boom period, mining continued in the district on a much reduced scale; all the known productive ground is said to have been worked over at least once. Most of the gold was recovered by first drywashing, then by wetwashing the dry-panned concentrates at the Gila River. A few large-scale operations have been attempted, but these were unsuccessful.

Source: Wilson (1933, p. 210) states that the gold in the Gila City placers probably came from many pockety or small low-grade gold veins in the northern end of the Gila Mountains. No high-grade gold veins are found in the vicinity of the placers.

Literature:

Browne, 1868: History of placer-mining activity.

Elliott, 1884: History—quotes Lieutenant Mowry's description of placers; placer-mining activity and production in 1858.

Farish, 1915a, v. 1: Repeats Lieutenant Mowry's description of placers.

Hinton, 1878: History of early placer-mining; quotes Browne's (1868) description of early history of Gila City.

Koschmann and Bergendahl, 1968: History; production.

Mowry, 1863: Early placer-mining activity.

Raymond, 1872: Extent; placer-mining operations.

———1874: Production information for 1873.

Trippel, 1889: Production statistics for 1888.

Wilson, 1933: History; location; production; bedrock geology; character of gravels; area of placer-mining activity; source of gold.
———1961: Virtually repeats Wilson (1933).

70. LAGUNA DISTRICT

Location: Laguna Mountains north of the Gila River, Tps. 7 and 8 S., Rs. 21 and 22 W.

Topographic map: Laguna Dam 7½ minute quadrangle.

Geologic map: Wilson, 1960, Geologic map of Yuma County, Arizona, scale 1:375,000.

Access: From Yuma, about 20 miles east on State Highway 95 to the Laguna Mountains; jeep trails and dirt roads lead to different placers in the mountains.

Extent: Three areas of placer concentration are known on the flanks of the Laguna Mountains. The McPhaul placer area is on the southern margin of the mountains; gravels have been drywashed from the Gila River to about 1¼ miles north of McPhaul Bridge (sec. 33, T. 7 S., R. 21 W.; sec. 4, T. 8 S., R. 21 W.). Also owned by H. H. McPhaul is the San Pablo placer claim of 160 acres, probably located in this same area. The Las Flores placer area is north of the McPhaul placer on the southeast slope of the Laguna Mountains. These small placers are found near the head of an alluvium-floored gulch in the vicinity of the old mines "Golden Queen and India" (abandoned mines located near the border of secs. 26 and 35, T. 7 S., R. 22 W.). Some gold was found in gravels in other gulches on the southern margin of the mountains as far south as the Gila River.

The Laguna Dam placer area is on the east side of the dam on the southwest flank of the Laguna Mountains. Gold was found in gulches draining the mountains, in the bed of the Colorado River, and in potholes in bedrock as high as 100 feet above the river. During the construction of the Laguna Dam in 1907, placer nuggets and a small gold-quartz vein was found at the river margin; part of the placer area was submerged after completion of the dam.

Production history: The placers at Laguna were worked about the time that the placers at Gila City were most active; early production is unknown. Production during the 20th century has been small and intermittent, and production is often grouped with production from the Gila City placers or given under the name "Colorado River placers."

Source: The placers in the three areas of the Laguna Mountains were derived from local gold-quartz veins in the metamorphic bedrock of the area.

Literature:
Koschmann and Bergendahl, 1968: History; placer-mining operations

in 1884 or 1885; production.

Mining Journal, 1941: Placer-mining operations at San Pablo placer.

Mining Review, 1910a: Notes submerging of placer ground in pothole area at Laguna Dam.

Raymond, 1872: History of placer-mining activity.

71. MUGGINS MOUNTAINS PLACERS

Location: In the Muggins Mountains, north of the Gila River, Tps. 7 and 8 S., Rs. 19 and 20 W.

Topographic maps: All 15-minute quadrangles—Wellton, Red Bluff Mountain, Laguna, Fortuna.

Geologic map: Wilson, 1960, Geologic map of Yuma County, Arizona, scale 1:375,000.

Access: From Yuma, 13 miles east on State Highway 95 to Blaisdell; 7 miles east on light-duty road to Dome. From there, dirt roads lead 10 miles northeast to the Muggins Mountains.

Extent: The only information I have found (other than production data) that describes the placers in the Muggins Mountains is that given by Wilson. The placers are found in the southern and central parts of the Muggins Mountains in the vicinity of Klothos Temple and Vinegaroon Wash.

Placers in the southern part of the range are in Burro Canyon (unlocated) and small canyons in the vicinity of Klothos Temple (sec. 1, T. 8 S., R. 20 W., Laguna quadrangle) and at the southern end of Long Mountain (sec. 7, T. 8 S., R. 19 W., Fortuna quadrangle). The gold-bearing gravels in Burro Canyon occur in ancient bars several feet above the stream channel and in the present stream channel. The gold occurs as particles as much as 0.15 inch in diameter and is concentrated at or near bedrock.

The placers in the central part of the range are near the headward forks of Vinegaroon Wash (which Wilson describes as a long northwestward-trending canyon that bisects the range) in T. 7 S., R. 19 W. (Laguna and Red Bluff Mountain quadrangles).

Production history: The placers in the Muggins Mountains have apparently been known for many years, but they have not been so actively worked as other placers in the same vicinity. They were probably most actively placered during the late 1800's; during the 20th century small-scale placer mining was carried on until 1942.

Source: The placer gravels in Burro Canyon are derived from a Miocene terrestrial conglomerate that contains detrital gold eroded from gold-bearing quartz veins in the crystalline rocks in the range. The minor placers in nearby small canyons and at the southern end of Long Mountain probably have a similar origin.

The gravels in Vinegaroon Wash, which reportedly yielded many rich pockets of gold, are thought to have been derived from erosion of the adjacent metamorphic rocks, which contain gold-bearing veins.

Literature:

Wilson, 1933: Location; source.

———1961: Location; extent; size and distribution of gold; source; placer-mining activity during the period 1932–33.

72. CASTLE DOME DISTRICT

Location: Southwest flank of the Castle Dome Mountains in the vicinity of Thumb Peak, Tps. 4 and 5 S., R. 18 W.

Topographic map: Castle Dome Mountains 15-minute quadrangle.

Geologic map: Wilson, 1960, Geologic map of Yuma County, scale 1:375,000.

Access: From Yuma, about 24 miles east and north on State Highway 95 to light-duty road leading east about 15 miles to jeep trail leading north along Big Eye Wash.

Extent: The placers in the Castle Dome district are reportedly located east and south of the Big Eye mine, a gold-quartz mine, at the head of Big Eye Wash, north of Thumb Peak (approximately sec. 34, T. 4 S., R. 18 W., unsurveyed). The gold is found near bedrock in gulches, but the exact location is unknown.

Production history: The placers are said to have been discovered in 1884, but the description of the discovery given equally fits the Tank Mountains placers (No. 73). Production during the 20th century was apparently small but nearly continuous until 1944. During the period 1940–42, a few hundred ounces of placer gold was recovered yearly from the Ocatilla placer (unlocated) by miners using a Stebbins dry concentrator.

Source: The gold is thought to be derived from the gold-quartz veins in the vicinity of the placers. The Castle Dome district is more famous for lead-silver veins near the Castle Dome mine (at the west side of the range) than for the gold-quartz veins, which are found in the eastern part of the range.

Literature:

Trippel, 1888: Notes placer activity.

U.S. Bureau of Mines, 1940–42: Reports operations at Ocatilla placer.

Wilson, 1933: General location; production.

———1961: General location; distribution of gold; source; production.

73. TANK MOUNTAINS PLACERS

Location: At the north end of the Tank Mountains near Engesser Pass and at the southeastern foothills of the Tank Mountains, T. 2 S., Rs. 15 and 16 W.; T. 4 S., R. 13 W.

Topographic maps: Engesser Pass and Palomas Mountains 15-minute quadrangles.

Geologic map: Wilson, 1960, Geologic map of Yuma County, Arizona, scale 1:375,000.

Access: From Yuma, 70 miles east on U.S. Highway 80 to Dateland, 9 miles north to light-duty road 3 miles south of Horn: from there, north on dirt roads 25 miles to the south edge of the Tank Mountains. Dirt roads lead into and around the mountains.

Extent: Placers are in many parts of the Tank Mountains and have been worked on a small scale since the 1870's. The only information I have found on the extent of these deposits is that given by Wilson, who describes two areas of placer concentration—near Engesser Pass and near Puzzles Mountain.

The Engesser placer is at the north end of the Tank Mountains (in the vicinity of the boundary between T. 2 S., R. 15 and 16 W., Engesser Pass quadrangle); the gold was recovered from gravels in the main gulch below the Engesser prospect (sometimes called the Johnnie Prospect) and from gravels in smaller nearby gulches.

The Puzzles area placer is in the southeastern foothills of the Tank Mountains, in the vicinity of a low ridge locally called Puzzles Mountains (NW¼ T. 4 S., R. 13 W.; Palomas Mountains quadrangle). The gold was recovered from shallow bench and stream gravels on the pediment near the Puzzles, Golden Harp, Ramey, and Regal prospects and is said to be coarser than the gold recovered near Engesser Pass.

Production history: The placers have apparently been worked intermittently on a very small scale since the 1870's, but because of the relative isolation of the district compared with those on the Gila River, very little information has been published. Wilson suggests that the placers in the Engesser mine area were probably worked earlier and with greater profit than the placers in the Puzzles area. Burchard describes the 1884 discovery of placers located 50 miles from Castle Dome Landing and 80 miles southeast of Ehrenberg; these deposits were probably found in the Tank Mountains, perhaps in the Engesser placer area. A small production of placer gold was reported from the Engesser area in 1936. The placers in the Puzzles area, which is 5 miles north of the Palomas Mountains, were actively drywashed in the early 1900's and probably produced placer gold attributed to the Palomas Mountains.

Source: The placers in the Tank Mountains were derived from local gold-bearing veins, which are mined at the Engesser, Puzzles, Golden Harp, Ramey, and Regal prospects.

Literature:
Burchard, 1884: New placer discovery; location; discoverer named; distribution of gold; production.
——1885: Repeats 1884 description; adds information on size and shape of gold; source.
Wilson, 1933: Location; history; source; bedrock geology.
——1961: Virtually repeats information in Wilson (1933).

74. KOFA DISTRICT

Location: Kofa Mountains, T. 2 S., R. 17 W.
Topographic map: Kofa Butte 15-minute quadrangle.
Geologic maps:
Jones, 1916, Geologic reconnaissance map of the Kofa Mountains, Arizona (pl. 5), scale 1:125,000.
Wilson, 1960, Geologic map of Yuma County, Arizona, scale 1:375,000.
Access: From Yuma, 59 miles east and north on State Highway 95 to dirt road leading east about 15 miles to dirt road leading about 10 miles north to the Kofa Mountains.
Extent: Many gulches in the southern and northeastern part of the Kofa Mountains are said to contain gold-bearing gravels; the only placer area described is in a westward-trending wash just north of the King of Arizona mine (secs. 1 and 12, T. 2 S., R. 17 W.). The gold-bearing debris consists of boulders and fragments of metamorphic and volcanic rocks and ranges in thickness from a few feet to 70 feet. The gold in the gravels is coarse and occurs near bedrock.

Very little information has been found relating to other placers in the Kofa Mountains. An underground placer mine, the Alamo mine, is said to be located in conglomerate and to have high gold values; the Alamo mine might be located near Alamo Spring in the Kofa Mountains (T. 1 N., R. 16 W.).
Production history: The only placer of importance in the area is the deposit north of the King of Arizona mine; the total production to 1914 was estimated at $40,000. During the 20th century small-scale intermittent placer-mining activity has produced small amounts of gold from this area. No production is known from the Alamo mine.
Source: According to Jones (1916), who studied the Kofa placer in detail, the source of the gold was small auriferous veins in the metamorphic rocks exposed north of the placer deposits in sec. 1, T. 2 S., R. 17 W., and not from the King of Arizona mine.
Literature:
Allen, 1922: Virtually repeats description of Jones (1916).
Jones, 1916: Location; production; placer-mining activity; thickness of gold-bearing gravel; size and distribution of gold; source.

McConnell, 1911: Alamo placer; gold values in conglomerate.

Wilson, 1933: Quotes Jones (1916), production from 1914 to 1928.

————1961: Quotes Jones (1916), placer-mining activity and production per man during the period 1932–33.

75. ELLSWORTH DISTRICT

Location: In the Granite Wash Mountains, southwest of the McMullen Valley, and in the Little Harquahala Mountains south of McMullen Valley, Tps. 5 and 6 N., R. 14 W.; T. 4 N., R. 13 W.

Topographic maps: Salome and Hope 15-minute quadrangles.

Geologic map: Wilson, 1960, Geologic map of Yuma County, Arizona, scale 1:375,000.

Access: From Yuma, 84 miles north on State Highway 95 to Quartzite; east 30 miles on U.S. Highway 60–70 to Hope. Dirt roads lead north about 5 miles to placer ground in the Granite Wash Mountains and southeast about 10 miles to vicinity of the Little Harquahala placers.

Extent: Gold-bearing gravels are found in many small gulches in the southeastern part of the Granite Wash Mountains and the southwestern part of the Little Harquahala Mountains.

Gulches near the Desert mine in the Granite Wash Mountains (sec. 21, T. 5 N., R. 14 W., unsurveyed, Salome quadrangle) were profitably drywashed about 1895. Placers were also found about 3 miles north of the Desert mine in the vicinity of the old Yellow Bird mine; gold-bearing gravels locally called Dutch Henry's diggings were located in the second wash south of the Yellow Bird Camp and west of the Arizona Northern prospect (secs. 32 and 33, T. 6 N., R. 14 W., unsurveyed, Salome quadrangle).

Placers in the Little Harquahala Mountains (T. 4 N., R. 13 W., Hope quadrangle) were worked from about 1884 to 1888 (at that time the area was known as the Centennial district). According to Burchard (1885), "Every gulch below the ore bodies (which include the Alps Group—unlocated) contains considerable placer gold which is flat but coarse"; Wilson (1961) reports that in 1886 and 1887 placers were worked in Harquahala Gulch in the vicinity of the Harquahala (Bonanza) mine, which was not discovered until 1888. In 1934 some placer gold was recovered from the Concepcion claim, 8 miles south of Wenden; although this claim is not identified on the topographic maps, it is probably located on the north flank of the mountains.

Production history: Early production from the placers in the Ellsworth district is only poorly known. One report indicates that 2 ounces per day was recovered for an unspecified short time from the placers near the Desert mine; gold worth $9,000 was recovered from these deposits about 1895. During 1888, $12,500 was recovered from the placers in the Harquahala Mountains (Centennial district).

During the first 30 years of the 20th century, the placers were intermittently mined on a small scale.

Source: The gold in the placers found in the Granite Wash Mountains was derived from small gold-quartz lenses in the schists of the area. Bancroft (1911, p. 98, 101) suggests that these veinlets formed after, but as a result of, the intrusion of the large granitic mass of Salome Peak.

The deposits in the Little Harquahala Mountains were probably derived by erosion of the Harquahala and similar veins in the area.

Literature:

Bancroft, 1911: Locates small placers in Granite Wash Mountains; history; production; source.

Burchard, 1885: Notes presence of placer gold.

Mining World, 1909: Notes approximate year of discovery; early placer production.

Trippel, 1889: Production statistics for 1888.

Wilson, 1961: Harquahala Gulch placer—location; history; production.

76. TRIGO DISTRICT (COLORADO RIVER PLACERS)

Location: West flank of the Trigo Mountains. T. 2 S., R. 23 W.; Tps. 3 and 4 S., Rs. 23 and 24 W.

Topographic maps: All 7½-minute quadrangles—Picacho NW, Picacho, Cibola.

Geologic map: Wilson, 1960, Geologic map of Yuma County, Arizona, scale 1:375,000.

Access: From Yuma, 71 miles north on State Highway 95 to Cibola Road, which leads 36 miles west to Cibola. One placer area is about 6 miles east of Cibola in the vicinity of the Hardt Gold mine, which is accessible by a dirt road. The Colorado River placers are about 13 miles south of Cibola and are accessible by a dirt road and jeep trial that parallels the river.

Extent: Small placers have been worked in drywashes in the Trigo Mountains for many years, but the location of these deposits is not described in the literature. Placers have been worked in drywashes south of the Hardt Gold mine in T. 2 S., R. 23 W., approximately secs. 1 and 2 (R. T. Moore, Arizona Bureau of Mines, oral commun., 1969).

Gold has been recovered from gravels along the west flank of the Trigo Mountains and possibly from Colorado River sands for many years. Very little information has been found about these deposits, which are located in the Paradise Valley section of the Colorado River Valley (Tps. 3 and 4 S., R. 24 W.).

Production history: Placers were mined in the Trigo Mountains as early as 1866, when placer activity was noted in the Silver subdistrict (T. 3 S., R. 23 W.). No estimate of early production is known. Placer-mining activity has been small scale throughout the 1900's and variously credited to the Trigo district and the Colorado River placers by the U.S. Bureau of Mines Minerals Yearbooks. Most of the placer gold was recovered from the Colorado River area, which was worked by many itinerant miners during the 1930's. From 1950 to 1966, these deposits were mined yearly by dry concentration at the "Colorado River Valley property"; descriptions of the deposit and production from 1960 to 1966 is held confidential by the U.S. Bureau of Mines.

Source: Small gold-bearing quartz veins in fault zones in schists in the Hardt mine area have been worked at the Hardt mine, the Boardway prospect, and the Jupiter claims (all in the Cibola subdistrict). The Boardway prospect and the Jupiter claim, located south of the Hardt mine, probably represent the type of gold deposit from which the placers were derived.

Literature:

Burchard, 1882: Notes placer activity in the Silver subdistrict in 1866.

U.S. Bureau of Mines, 1950, 1951, 1953: Reports production from the "Colorado River Valley Property."

Wilson, 1961: Reports placer mining in Trigo Mountains; distribution and shape of gold (location is in error owing to lack of surveyed townships and ranges at the time of writing, about 1933).

Wilson, 1933: Describes lode mines in Cibola region and other mines in Trigo Mountains.

77. LA PAZ DISTRICT

Location: West flank of the Dome Rock Mountains, Tps. 3 and 4 N., Rs. 21 and 22 W.

Topographic maps: Dome Rock Mountains 15-minute quadrangle; La Paz Mountain 7½-minute quadrangle.

Geologic maps:

Jones, 1915, Map showing geology of southern part of Colorado River Indian Reservation and location of placers near Quartzsite, Arizona (pl. 4), scale approximately 3 miles = 1 in.

Wilson, 1960, Geologic map of Yuma County, Arizona, scale 1:375,000.

Access: From Yuma, 84 miles north on State Highway 95 to Quartzsite, about 10 miles west on U.S. Highway 60–70 to placers, which are adjacent to the highway and in nearby gullies and washes.

Extent: Placers in the La Paz district are found in Goodman Arroyo and Arroyo La Paz, major west-trending drainages, and in Ferrar, Garcia, and Ravenna Gulches, tributaries to the major drainages. Placers were worked as far west as the outskirts of the town of Ehrenberg (secs. 15 and 16, T. 3 N., R. 22 W.).

The gold-bearing gravels range in thickness from a few feet on the mountain slopes to an unknown depth in Arroyo, La Paz, and Gonzales Wash (the wash through which U.S. Highway 60–70 is built); gold is distributed throughout the gravels, but the richest parts were found near bedrock. Ferrar Gulch (secs. 25 and 36, T. 4 N., R. 21 W.) reportedly contained the richest gravels in the area, and it was from this gulch that Juan Ferrar recovered a nugget weighing more than 47 ounces.

Production history: The placers in the La Paz district were discovered by Captain Pauline Weaver in January 1862, when he panned a small amount of gold from a gulch called El Arollo de la Tenaja in the Dome Rock Mountains. Immediately thereafter, Weaver returned to Yuma, told about his discovery, and on his return to the mountains was joined and followed by other prospecting parties (Browne, 1868, p. 454). According to Browne, this advance party soon found good prospects (one man, Jose Redondo, recovered a nugget weighing more than 2 oz in a place less than a mile south of Weaver's camp) and returned to Laguna for supplies. The real rush to the new placer soon followed. About $1 million in placer gold was recovered from the placers the first year and another $1 million by 1864, when the placers were worked out. Since that time, the La Paz district has been at times part of the Colorado River Indian Reservation, and in consequence, small-scale placer-mining activity declined and large-scale placer-mining plans were interrupted. During the 20th century, while many large-scale operations were active in the Plomosa district to the east, the La Paz placers were worked only by individuals.

The La Paz placers are famous for the large nuggets recovered, although most of the gold occurred as pieces ranging in weight from 0.0025 to 0.5 ounce. Browne (1868, p. 454) describes large nuggets from the gravels, some weighing 26, 27, and 47 ounces, that were free of all foreign material, even quartz, and thought it possible that many larger nuggets were recovered but not shown for various practical or superstitious reasons. The largest nugget recovered from the La Paz placers was valued at about $1,150 (about 65 oz) and assayed 870 fine (Heikes and Yale, 1913, p. 259; Jones, 1915,

p. 50); Jones attributes the recovery of this nugget to Juan Ferrar. As late as the 1930's, a nugget valued at $900 (about 45 oz) was recovered by a prospector (MacHunter and Henderson, 1958); most of the large nuggets were found years earlier; later workers found only smaller particles.

Source: The discovery of the La Paz placers led to intensive prospecting for large gold-bearing veins, but major lode mines have not been developed. The gold in the placers is attributed to erosion of the many gold-bearing veins distributed through the metamorphic rocks in the area. Some of these veins, such as the Goodman vein, which trends northwest-southeast across the north end of Ferrar Gulch and is exposed in Gonzales Wash, have been mined in the past. Jones (1915, p. 54) states that the decomposition of these veins has produced the placer gold, as the largest areas of placer gravels are found along the more persistent gold-quartz veins.

Literature:

Allen, 1922: Location; quotes description of Jones (1915); placer-mining operations; gold values per cubic yard; production.

Bancroft, 1911: Quotes Browne (1868).

Browne, 1868: Gives a complete history of discovery and early activity in placers in the La Paz district on p. 454–455.

Heikes and Yale, 1913: Location; thickness of gravels; gold values per cubic yard; size of nuggets; placer-mining operations.

Hinton, 1878: History of early placer mining; later placer discoveries.

Jones, 1915: Detailed description of La Paz placers; mining methods; depth and extent of workings; amount and type of gold removed. Relations to local geology discussed.

Koschmann and Bergendahl, 1968: History; production.

MacHunter and Henderson, 1958: Popular account of discovery and subsequent development of La Paz diggings. Photographs of area included.

Randolph, 1901: Size of nuggets recovered.

Raymond, 1872: Geology; origin of the placers.

———1874: Production information for 1873.

Trippel, 1889: Production statistics for 1888.

Wilson, 1961: History; location; placer-mining techniques; source; quotes descriptions of earlier studies.

78. LA CHOLLA, MIDDLE CAMP, AND ORO FINO PLACERS

Location: East side of the Dome Rock Mountains, Tps. 3 and 4 N., R. 20 W.

Topographic map: Dome Rock Mountains 15-minute quadrangle.

Geologic map: Wilson, 1960, Geologic map of Yuma County, Arizona, scale 1:375,000.

Access: From Quartzsite it is about 5 miles west on U.S. Highway 60–70 to Dome Rock Mountains. Dirt roads lead from the highway to adjacent placers.

Extent: Placers are found on the east side of the Dome Rock Mountains in three localities: La Cholla placers, located south of U.S. Highway 60–70 in an area 4 to 5 miles long from east to west (NE¼ T. 3 N., R. 20 W.); at the southern base of the Middle Camp Mountain (approximately SW¼ T. 4 N., R. 20 W., unsurveyed); Oro Fino placers, located north of U.S. Highway 60–70 and southeast of the Middle Camp placers (SE¼ T. 4 N., R. 20 W., unsurveyed).

Production history: The placers on the east side of the Dome Rock Mountains have been worked intermittently since the 1860's. The placer camps are usually considered to be part of the Plomosa mining district and production records are usually grouped with the Plomosa placers, 12 miles east. The deposits at La Cholla and Middle Camp have been worked on a small scale throughout this century. From the 1930's until 1941, the La Cholla placers were the most active in the area; large-scale operations by the La Posa Development Co. at the Arizona Drift mine during the period 1939–41 accounted for most of the placer production. This company worked a channel on bedrock that was 6 feet thick and 50–150 feet wide at a depth of 140 feet. No estimates of the average value of the gravels worked by this operation have been found, but the gravels worked in an earlier operation contained in places as much as 1 ounce of gold per cubic yard, and the tailings were valued at 85 cents per cubic yard.

Source: Gold-quartz veins in the mountains immediately adjoining the placer deposits are the source of the placer gold.

Literature:

Allen, 1922: Quotes Heikes and Yale (1913).

Berger, 1932: Placer-mining techniques and operations at La Cholla.

Gardner and Johnson, 1935: Depth of gravels; width of gold-bearing channel; placer-mining operations.

Heikes and Yale, 1913: Location; types of gravel at different localities; size of nuggets; past placer-mining history.

Jones, 1915: Placer-mining operations at Oro Fino; gold values in gravels.

Mining Journal, 1939a: Operations by La Posa Development Co., number of men employed; yards of gravel mined per day; equipment used; thickness of pay gravel; width of channel.

U.S. Bureau of Mines, 1940: Reports cubic yards of gravel mined at La Cholla placers in 1939.

Wilson, 1961: Location; placer-mining activity; distribution of gold-bearing gravels.

79. PLOMOSA DISTRICTS

Location: West side of the Plomosa Mountains, Tps. 3 and 4 N., R. 18 W.; T. 2 N., R. 17 W.

Topographic map: Quartzsite 15-minute quadrangle.

Geologic map: Miller, 1970, Geologic map of the Quartzsite quadrangle, Yuma County, Arizona, scale 1:62,500.

Access: From Quartzsite, it is about 5 miles southeast on dirt roads to placer area.

Extent: The Plomosa placers are found at the western edge of the Plomosa Mountains in and near Plomosa Wash south of Scaddan Mountain (Tps. 3 and 4 N., R. 18 W.). The extent of the placers in the Plomosa district is not known, but some reports indicate that gold-bearing gravels are found for about 3–4 miles along the western edge of the mountains. Most large-scale placer activity was apparently concentrated in secs. 3 and 4, T. 3 N., R. 18 W., near the old town of Plomosa. Small placers were apparently worked in Cave Creek in the southern part of the range (T. 2 N., R. 17 W.).

Production history: The Plomosa placers were worked in the early 1860's about the same time as the La Paz placers. Early production is unknown, but the extensive diggings remaining from the early workers indicate that production was large. The Plomosa placers were actively sampled and mined during the 1910's by companies using large-scale drywashing machines; these operations attracted much attention from mining men at the time, and the resulting literature described the gravels and mining techniques in great detail. Production records indicate that these operations were not a commercial success. Various reports of the average value of the gravels claim values as high as $20 per yard in placer gold, but more reasonable estimates range between 71 cents to $1.77 per yard. The placers worked by the Yuma Consolidated Co. and the Plomosa Placer Co. during the period 1915–16 were cemented gravels as thick as 140 feet; the gold was concentrated at various horizons in the gravels. The placers were steadily mined by individual drywashers until the 1950's.

Source: Gold-bearing veins occur in the metamorphic rocks in the immediate vicinity of the placers and are considered to be the source of the gold in the gravels.

Literature:

Allen, 1922: Quotes Heikes and Yale (1913).

Bancroft, 1911: Placer-mining techniques and operations; character of the placer gravels; source.

Heikes and Yale, 1913: Location; character and thickness of placer gravel; gold values per cubic yard; results of sampling gravels; placer-mining operations; information is abstracted from a professional report (unpublished) by John A. Church; production from 1904 to 1912.

Jones, 1915: Quotes Bancroft (1911); placer-mining activity in 1914.

Keiser, 1916: Describes placer-mining operations and techniques used by Yuma Consolidated and Plomosa Placer Cos.

Koschmann and Bergendahl, 1968: History; production.

Maltman, 1917: Placer-mining techniques and operations; character, thickness of average value of placer ground.

Mining and Scientific Press, 1916: Note on placer plants operated by Yuma Consolidated and Plomosa Placer Cos. Includes average value of different gravel horizons in 140-foot-deep hole on Yuma Consolidated land.

Plummer, 1916: Placer-mining techniques and operations by Yuma Consolidated Co. and Plomosa Placer properties; character, thickness, average value of placer gravels; distribution of gold.

Root, 1912: Gold values in gravels.

Wilson, 1961: Repeats information given by Bancroft (1911) and Heikes and Yale (1913); placer-mining activity during the period 1932–33.

OTHER DISTRICTS

80. CIENGA DISTRICT

Placer gold was recovered in 1910 from this district, which is in the Buckskin Mountains on the east side of the Colorado River just south of the junction with the Bill Williams River (T. 10 N., R. 18 W.). The district contains small copper and gold deposits consisting of small pockets of ore minerals in shear zones in sedimentary rocks; some pockets were rich in free gold.

Literature:

Wilson, Cunningham, and Butler, 1934: Describes lode deposits.

81. COCOPAH DISTRICT

Placer gold was credited to this district in 1934. I have found no information about a district of this name.

82. FORTUNA DISTRICT

The Fortuna district is on the west flank of the Gila Mountains at the old Fortuna mine (T. 10 S., R. 20 W.). The Fortuna mine was actively worked between 1896 and 1904, producing more than $2.5 million in gold. Most of the gold credited to placer production from this area was actually recovered by placer-mining techniques used to rework old tailings at the mine.

Literature:

Wilson, Cunningham, and Butler, 1934: Describes Fortuna mine.

83. LA POSA DISTRICT

This district is in the Wellton Hills (T. 10 S., R. 18 W.), a small range between the Gila Mountains and the Copper Mountains. Many low-grade gold-quartz veins occur in the hills and have been mined on a small scale. Placer gold was recovered in the early 1900's by intermittent prospectors.

Literature:

Wilson, 1933: Describes district.

84. MOHAWK DISTRICT

One ounce of placer gold was credited to Mohawk in 1940, but no information has been found about the source of the gold. The gold may have been recovered elsewhere and sold for supplies in Mohawk, a station on the Southern Pacific Railroad.

85. SANTA MARIA DISTRICT

Placer gold was recovered from this district, also known as the Planet district, in 1912 and between 1959 and 1963. Some of the placer production is confidential. The Santa Maria district is in the Buckskin Mountains, on the south side of the Bill Williams River. The district is noted for copper deposits, but some gold ores do occur. Before 1911 the Planet Copper mine controlled three placer claims in that area; their location is not described in the literature.

Literature:

Bancroft, 1911: Notes placer claims in area.

Wilson, Cunningham, and Butler, 1934: Describes gold prospects worked during the period 1933–34.

86. SHEEP TANKS DISTRICT

Placer gold was recovered from this district in the Little Horn Mountains in 1936 and 1940. Most of the gold mineralization in the mountains occurs in the area of the Sheep Tanks mine (approximately sec. 1, T. 1 S., R. 15 W.), and the placer gold was probably recovered from small deposits in that area.

Literature:

Wilson, 1933: Describes Little Horn Mountains and Sheep Tanks mine.

87. SONORA DISTRICT

Three ounces of placer gold was credited to Sonora in 1935, but no information has been found about the source of the gold.

GOLD PRODUCTION FROM PLACER DEPOSITS

Arizona ranks tenth in the United States (eighth in the western continental States) in placer gold production. The U.S. Bureau of Mines (1967, p. 15) cites 500,000 troy ounces of placer gold produced in Arizona from 1792 to 1964. I estimate that placer gold production was at least 564,052 ounces. Districts of largest placer production were the Lynx Creek, Big Bug, and Weaver (Rich Hill) districts (Yavapai County), the Gila City (Dome), and La Paz district (Yuma County), and the Greaterville district (Pima County), all with estimated placer production of more than 25,000 ounces. The available production information for all placer districts is given in table 1. For comparison, I have included table 2, which lists 27 gold districts in Arizona that have produced more than 25,000 ounces of gold (from Koschmann and Bergendahl, 1968).

As table 1 clearly shows, Arizona has many small placer-mining districts from which only a few ounces of gold has been recovered, mostly during the depression years of the 1930's. For most of these districts, little information other than production has been found. Comparing table 1 with table 2 shows that the major lode-gold districts in the State, except for the Bradshaw Mountains in Yavapai County, have had very little placer gold production.

Most of the placer gold produced in the State of Arizona was recovered by tedious work on a small scale by individuals who used rockers, pans, sluices, and dry concentrators. In only a few districts have large-scale placer-mining operations been successful, although many attempts were made to use large dry-concentrating machines. The most successful large-scale operations have been in the Lynx Creek and the Big Bug districts, Yavapai County, where the presence of adequate supplies of water enabled large dredges to mine the gold-bearing gravels. Among the largest and most profitable large-scale dry-concentrating operations were those in the San Domingo Wash district, Maricopa County, in the Plomosa district, and at La Cholla placers, Yuma County; at Copper Basin, Yavapai County, the gravel was hauled to a central washing plant where wet methods of recovery were used.

TABLE 1.—*Arizona placer gold production, in ounces*

[Data in brackets are ounces of gold reported by the U.S. Bureau of Mines for districts in drainage areas]

Map locality (pl. 1)	County and placer district	Estimated production, discovery to 1900	Recorded production data from U.S. Bur. Mines			Total recorded production 1902–68	Total estimated production	Reference source for estimated production
			1902–33	1934–42	1943–68			
	Cochise:							
1	Dos Cabezas-Teviston	0	474	464	7	945	>1,000	
2	Courtland-Gleeson	0	46	6	0	>52	100	
3	Bisbee-Warren	0	4	271	0	275	>275	
4	Huachuca	Unknown	141	58	2	201	>200	
5	California	0	116	0	0	116	116	
6	Pearce	Unknown	0	0	0	0	≈250	
7	Rucker Basin	0	0	0	1	1	1	
	Gila:							
8	Payson (Green Valley)	Unknown	0	44	5	49	100	
9	Globe-Miami	550	182	123	5	310	1,000	Hinton (1878); Trippel (1888, 1889).
10	Barbarossa-Dripping Spring	0	38	98	1	137	500	
11	Mazatzal Mountains	0	18	0	0	18	18	
12	Young	0	13	0	0	13	13	
	Graham and Greenlee:							
13	Clifton-Morenci	2,500	112	78	0	190	2,700	Lindgren (1905a).
14	San Francisco River placers	---	772	192	8	972	3,000	
15	Gila River placers	---	13	1	0	14	20	
16	Rattlesnake	Unknown	5	0	0	5	5	
	Maricopa:							
17	Big Horn	0	>38	173	1	>212	>250	
18	Vulture	Unknown	134	162	12	308	>500	
19	San Domingo	Unknown	998	467	85	1,550	>2,000	Trippel (1889).
20	Cave Creek	>363	149	15	1	165	>528	
21	Aqua Fria	0	41	2	0	43	50	
22	Dads Creek	0	0	3	0	3	3	
23	Eagle Tail Mountains	0	2	0	0	2	2	

No.	Locality							Reference
24	New River	0	4	0	0	4	4	
25	Pikes Peak	0	0	13	1	14	>14	
26	Sunflower	0	0	2	0	2	2	
	Mohave:							
27	Chemehuevis	Unknown	60	202	2	264	1,000	
28	San Francisco	Unknown	48	7	0	55	100	
29	Kingman area placers	Unknown	22	1	0	23	50	
30	Colorado River placers	Unknown	10	17	0	27	50	
31	Gold Basin	0	140	413	13	566	1,000	
32	Lost Basin	0	130	672	>22	824	1,000	
33	Cotonwood	0	14	0	0	14	14	
34	Owens	Unknown	10	0	0	10	10	
	Wallapai	0	0	<1	0	<1	1	
	Pima:							
35	Alder Canyon placers	0	0	12	0	12	50	
36	Greaterville	32,500	3,522	347	543	4,412	40,000	Elsing and Heineman (1936).
37	Arivaca	7,500	80	245	8	333	10,000	Willis (1915).
38	Sierrita Mountains	12,500	7	2	0	9	12,500	Elsing and Heineman (1936).
39	Baboquivari	0	0	16	12	28	50	
40	Cababi	Unknown	663	6	0	669	700	Wilson (1961).
41	Quijotoa	12,500	1,775	121	2	1,898	15,000	
42	Ajo	0	24	0	0	24	24	
43	Empire	0	0	2	0	2	2	
44	Old Baldy	Unknown	0	0	0	0	Unknown	
45	Silver Bell	0	9	0	0	9	9	
	Pinal:							
46	Old Hat	Unknown	545	228	2	775	>1,000	
47	Casagrande	0	5	0	0	5	5	
48	Goldfield	0	1	0	2	3	3	
49	Mineral Creek (Ray)	20	0	0	0	0	20	Trippel (1889).
50	Pioneer (Superior)	0	0	3	0	3	3	
	Santa Cruz:							
51	Oro Blanco	1,000	77	26	0	103	1,500	Koschmann and Bergendahl (1968).

TABLE 1.—Arizona placer gold production, in ounces—Continued

Map locality (pl. 1)	County and placer district	Estimated production, discovery to 1900	Recorded production, data from U.S. Bur. Mines			Total recorded production 1902-68	Total estimated production	Reference source for estimated production
			1902-33	1934-42	1943-68			
	Santa Cruz—Continued:							
52	Nogales	Unknown	15	6	0	21	50	Elsing and Heineman (1936).
53	Patagonia	Unknown	79	25	0	104	104	
54	Tyndall-Palmetto-Harshaw	Unknown	6	0	0	6	50	
	Yavapai:							
55	Lynx Creek drainage area	51,500	4,060	24,390	567	29,017	80,000	
	Lower Lynx Creek	---	---	[24,155]	[543]	---	---	
	Upper Lynx Creek	---	---	[234]	[24]	---	---	
56	Hassayampa River drainage area.	25,000	1,162	1,958	281	3,401	30,000	Do.
	Upper Hassayampa River	[17,500]	[643]	[1,542]	[236]	---	---	
	Central Hassayampa River	Unknown	[10]	[25]	[4]	---	---	
	Walnut Grove	[2,500]	[344]	[257]	[7]	---	---	
	Minnehaha, Cherry, and Oak Creeks.	[5,000]	[137]	[36]	[1]	---	---	
	Lower Hassayampa River	Unknown	[28]	[98]	[33]	---	---	
57	Big Bug Creek drainage area	30,000	3,978	11,916	1,369	17,263	50,000	Do.
58	Turkey Creek drainage area	Unknown	92	90	4	186	1,000	
59	Black Canyon drainage area	2,500	82	341	27	450	3,000	
60	Humbug Creek drainage area	Unknown	395	701	21	1,117	1,500	
	Castle Creek	---	[40]	[157]	[12]	---	---	
	Humbug Creek	---	[233]	[494]	[7]	---	---	
	Pine Grove	---	0	0	0	---	---	
	Silver Mountain	---	[38]	[27]	[1]	---	---	
	Tip Top	---	[73]	[22]	[0]	---	---	
	White Picacho	---	[11]	[1]	[0]	---	---	
61	Weaver district	100.000	5,689	1,797	74	7,560	110,000	Do.
	Weaver	[100,000]	[5,679]	[1,720]	[74]	---	---	
	Martinez	---	[10]	[77]	[0]	---	---	
62	Model and Kirkland	---	125	34	5	164	200	

	District							
63	Copper Basin	5,000	1,369	1,055	7	2,431	8,000	Do.
64	Granite Creek	>1,000	103	43	4	150	1,500	Do.
65	Eureka	Unknown	85	92	22	199	1,000	
66	Lincoln Creek (unlocated)	0	19	0	0	19	19	
67	Pocket Creek (unlocated)	0	0	201	179	380	380	
68	Squaw Creek	0	7	0	0	7	7	
	Yuma:							
69	Gila City	25,000	240	344	29	613	>26,000	Wilson (1961).
70	Laguna	Unknown	1,202	439	0	1,641	5,000	
71	Muggins Mountains	Unknown	44	196	0	240	>500	
72	Castle Dome	5,000	1,108	836	9	1,953	>7,000	Do.
73	Tank Mountains	Unknown	111	4	0	115	200	
74	Kofa	<2,000	1,547	39	7	1,593	2,500	Do.
75	Ellsworth	1,000	264	2	0	>266	1,500	Bancroft (1911); Trippel (1889).
76	Trigo	0	274	150	>626	1,050	1,500	Wilson (1961).
77	La Paz	100,000	412	1,075	26	1,513	>100,000	Do.
78	La Cholla, Middle Camp, and Oro Fino placers.	12,500	1,685	3,109	18	4,812	18,000	Do.
79	Plomosa	12,500	1,682	1,792	>203	3,677	18,000	Do.
80	Cienega	0	5	0	0	5	5	
81	Cocapah (unlocated)	0	0	1	0	1	1	
82	Fortuna	0	36	23	0	59	59	
83	La Posa	0	27	0	0	27	27	
84	Mohawk	0	0	1	0	1	1	
85	Santa Maria	0	2	0	>99	>101	>200	
86	Sheep Tanks	0	0	4	0	4	4	
87	Sonora	0	0	3	0	3	3	
	Total		36,347	55,160	4,312	95,819		
	Undistributed		143	269	204	616		
	State total	442,433	36,490	55,429	4,516	96,435	564,052	

TABLE 2.—*Arizona lode-gold production, in ounces*

[Districts having production greater than 25,000 oz]

County and district	Production
Cochise County:	
1 ____ Bisbee	2,193,000
2 ____ Tombstone	271,200
3 ____ Turquoise	70,000
Gila County:	
4 ____ Banner	26,000
5 ____ Globe-Miami	191,801
Graham and Greenlee Counties:	
6 ____ Clifton-Morenci	203,000
Maricopa County:	
7 ____ Vulture	366,000
Mohave County:	
8 ____ San Francisco	2,045,400
9 ____ Wallapai	125,063
10 ____ Weaver	63,200
Pima County:	
11 __ Ajo	990,000
Pinal County:	
12 ____ Mammoth	403,000
13 ____ Ray	35,250
14 ____ Superior	398,200
Santa Cruz County:	
15 ____ Oro Blanco	100,200
Yavapai County:	
16 ____ Big Bug	584,300
17 ____ Black Canyon	46,700
18 ____ Eureka	59,966
19 ____ Hassayampa-Groom Creek	108,300
20 ____ Jerome	1,571,000
21 ____ Lynx Creek-Walker	43,000
22 ____ Martinez	396,300
23 ____ Pine Grove-Tiger	130.275
24 ____ Weaver-Rich Hill	204,000
25 ____ Ellsworth	134,000
26 ____ Fortuna	125,332
27 ____ Kofa	237,000

The total amount of placer gold recovered yearly in Arizona from 1900 to 1968 is graphed in figure 1, which also shows major contributors to the peak production.

SUMMARY

The ultimate source of detrital gold in placer deposits is, for the most part, gold-bearing lode deposits, which in Arizona are represented by veins in faults, fissures, and shear zones of various sizes.

Most of the placer gold found in Arizona was derived from systems of small gold-quartz veinlets and stringers scattered throughout the bedrock of the adjacent mountain ranges; in only a few localities

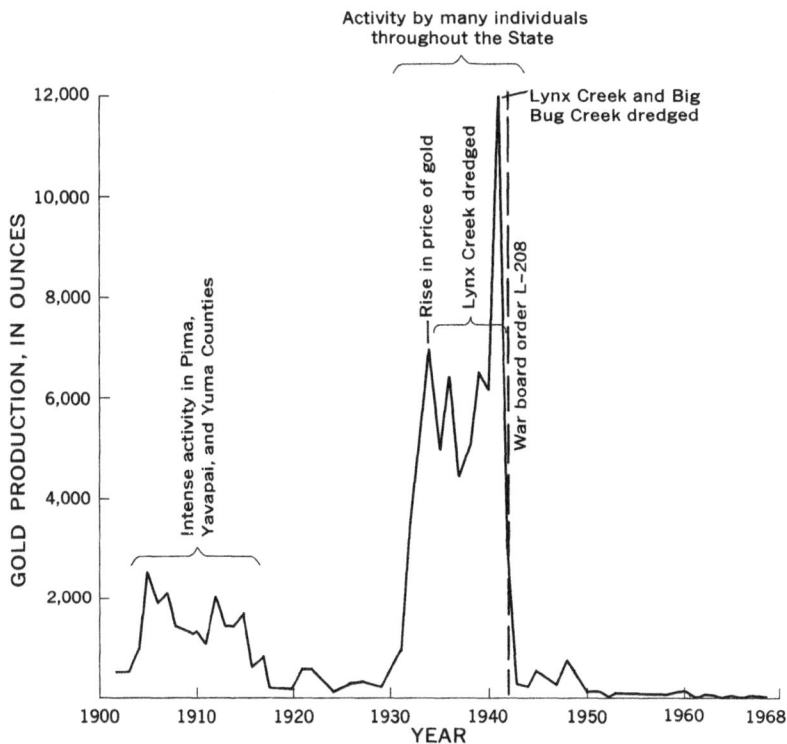

FIGURE 1.—Arizona placer gold production, in ounces.

was the gold in large placer deposits derived from vein systems of sufficient size to encourage lode mining on a large scale. Small placers commonly occur near large gold lodes, but are generally not economic. The most productive gold veins are those formed during Laramide time, which occur in rocks of Precambrian to Laramide (Late Cretaceous and early Tertiary) age. Much gold has been recovered as a byproduct from copper and other base-metal ores. Since 1941 the large copper mines have been predominant in the production of lode gold (Wilson, 1962).

RELATION BETWEEN PLACER DEPOSITS AND ROCK DEPOSITS

One of the outstanding characteristics of placer-mining history in Arizona is the large number of mining districts in which placer mining, albeit some minor, was carried on. A comparison of the placer gold production (table 1) with the major lode-gold production (table 2) shows that only the Tombstone and Jerome districts had no placer production at all. At Tombstone, the gold occurs as native gold in very fine particles; at Jerome, the gold was recovered from copper ores as a byproduct. In the mining districts that had a very low production of placer gold, the gold occurred as very fine particles, commonly included within copper and other base-metal ores, and therefore was not concentrated during the erosional cycle (see Koschmann and Bergendahl, 1968, p. 32–53, for a summary of gold characteristics in major lode-gold districts in Arizona).

Most placer deposits in Arizona are found in gullies and arroyos developed on alluvial fans or pediments on flanks of mountain ranges; a few placers occur as small local concentrations developed in gulch and hillside gravels immediately adjacent to the lode source of the gold. Wilson (in Webber, 1935) asserts that many of southwestern Arizona's largest and richest placers were derived from the erosion of numerous auriferous veinlets or stringers. He states

Ten one-quarter inch stringers with an average of 0.50 of gold per ton through a length of 500 feet and a depth of 300 feet could yield $24,000 worth of placer gold (with gold at $20.67 per ounce) while a million dollar placer (with the same gold value) could result from the erosion of 416 stringer bodies of that same very modest width, length, depth, and tenor, provided perfect concentration obtained.

Typical of the placers formed on extensive alluvial fans or on pediments are the San Domingo Wash placers, Maricopa County; the Lost Basin-Gold Basin placers, Mohave County; the Cañada del Oro placers, Pinal County; Rich Hill placers and Copper Basin placers, Yavapai County; the Quijotoa, Arivaca, and Greaterville placers, Pima County; the Laguna, Gila, and La Paz placers, Yuma County; Dos Cabezos placers, Cochise County. The gold-bearing gravels occur on hillslopes and in gullies of present-day and older drainage systems; the gold is generally concentrated on a false bedrock of caliche-cemented gravels but occurs in smaller amounts throughout the gravel.

The placer deposits in the Bradshaw Mountains, Yavapai County, are, as a group, the most productive and extensive placers in the State. The Bradshaw Mountains are also the locale of many productive gold lode deposits (table 2). The placers were productive because they were derived from large systems of well-developed gold-quartz veins of Laramide and Precambrian ages. They were extensive because of the wide distribution of the veins and the vast number of valleys and gulches that drain the comparatively well watered Bradshaw Mountains.

Placers found in other, comparatively well watered mountain ranges (Globe-Miami district, Gila County; Clifton-Morenci, San Francisco

River districts, Greenlee County), are similar in depositional features to the placers in the Bradshaw Mountains, but not in quantity of gold.

Typical of placers occurring as small local concentrations of gold in gravels immediately adjacent to the lode source of the gold are the Maud Hill placers, Courtland-Gleeson district, Gold Gulch placers, Bisbee-Warren district, Cochise County; Payson placers, Gila County; Vulture placers, Maricopa County; Oatman placers, Mohave County; Ajo placers, Sierrita placers, Pima County; Eureka placer, Yavapai County; Kofa, Harquahala, and Tank Mountains, Yuma County.

AGE OF PLACER GRAVELS

All economic placer deposits in the State were formed during the Quaternary Period. A few uneconomic deposits in the Bradshaw Mountains, Yavapai County (Milk Creek-Slate Creek drainage, and possibly some deposits along the Hassayampa River drainage), are found in Tertiary sedimentary rocks. The placer deposits in part of the Muggins Mountains, Yuma County, and in the Lost Basin district, Mohave County, were apparently derived from low-grade placer deposits of late Tertiary age. In only a few places can a maximum age for the gold-bearing gravels be estimated. At Gila City, Laguna, and the Muggins Mountains, Cretaceous and Tertiary sedimentary rocks form the bedrock of the placers; at Greaterville, parts of a bison skeleton considered to be Pleistocene in age was found in a placer deposit. It appears from a study of the literature that the gold-bearing alluvial deposits of the Bradshaw Mountains are products of the present-day drainage system, but a number of small alluvial flats in the mountains that may represent earlier erosional cycles may contain placer gold.

AGE OF LODE MINERALIZATION

The importance of mineralization associated with the Laramide orogeny has been emphasized by many geologists who study Arizona ore deposits. Most of these studies, and most geochronologic studies of ore-bearing rocks, have concentrated on districts in the central and southeastern parts of the State. Little is known about the age of mineralization of the western and southwestern parts of the State, as few detailed geologic studies have been made of the ore-bearing rocks of that region.

Most of the placer gold recovered from deposits in the Bradshaw Mountains is thought to be derived from Laramide veins, which are more prevalent and contained greater amounts of gold than the Precambrian veins. Although most evidence for the age of veins is based on observations of geologic relations between veins and country rock, two potassium-argon dates (see Anderson, 1968, p. 1169) on granodiorites considered to be genetically related to ore deposits demonstrate a **Laramide age (64 m.y., 70 m.y.)** for the veins in the Walker and Big Bug districts.

A Laramide age is known or assumed for the following other important lode deposits from which placer deposits have been derived: Maricopa County—Vulture vein, Vulture district; Yavapai County—Hillside vein, Eureka district; Octave vein, Weaver district; Yuma County—Harquahala vein, Ellsworth district; Fortuna vein, Fortuna district. A Laramide age is known or assumed for the following areas in which placer deposits derived from small vein systems have been found: Cochise County—Gold Gulch, Bisbee-Warren district; Greenlee County—small gold veins, Clifton-Morenci district; Pima County—gold veins in Cretaceous sedimentary rocks, Greaterville district; Baboquivari district; Arivaca district; Quijotoa district; Ajo placer, Ajo district; Yuma County—Plomosa area, Plomosa district; Kofa placers, Kofa district (Kofa vein is assumed to be Laramide, but placers were not derived from that vein).

Placer deposits that have been derived from lodes found in Precambrian rocks and that may be Precambrian in age in addition to those found in the Bradshaw Mountains, include those in the Globe district and the Payson district, Gila County; Lost Basin-Gold Basin districts and Chemehuevis district, Mohave County. Many placer deposits in Yuma County are derived from small gold veinlets in metamorphic rocks that have been considered to be Mesozoic or Precambrian in age. Detailed study in one area of Yuma County (the Plomosa Mountains, Miller, 1970) indicates that the metamorphic rocks exposed are of both ages, but until further study is made of this county, the age of mineralization can only be considered unknown.

Two areas in the State contain placer deposits derived from lode deposits of middle and late Tertiary age (post-Laramide). Some gold veins in the Dos Cabezas Mountains, Cochise County, have been dated at younger than 29 m.y. The most productive gold veins in the State, the veins in the San Francisco district, Mohave County, were the source of only small amounts of placer gold and are considered to be late Tertiary in age.

BIBLIOGRAPHY

LITERATURE REFERENCES

Allen, M. A., 1922, Arizona gold placers: Arizona Bur. Mines Bull. 118, 24 p.
Twenty-two placer areas are described in this first bulletin of the series in gold placers in Arizona.

————1923, History and present status of Arizona's gold placers: Mining Jour. [Phoenix, Ariz.], v. 7, no. 2, p. 5–7, 40; v. 7, no. 3, p. 6–7, 25.
Article extracted from Bull. 118.

Anderson, C. A., 1968, Arizona and adjacent New Mexico, in Ridge, J. D., ed., Ore deposits of the United States, 1933–1967 (Grafton-Sales volume): New York, Am. Inst. Mining, Metall., and Petroleum Engineers, Inc., v. 2, p. 1163–1190.

————1972, Precambrian rocks in the Cordes area [Turkey Creek placer], Yavapai County, Arizona: U.S. Geol. Survey Bull. 1345, (in press).

Arizona Bureau of Mines, 1938, Some Arizona ore deposits: Arizona Bur. Mines Bull. 145, 136 p.
Collection of papers detailing some mining districts in the State. Papers originally presented at American Institute of Mining and Metallurgy Engineers meeting in Tucson, November 1938.

Arizona Engineer and Scientist, 1961, Engineers create possiblie Arizona Gold Rush with "Rube Goldberg" dredge: Arizona Engineer and Scientists, v. 4, no. 10, p. 8–10.

Arizona Mining Journal, 1919, Placer mining to the fore [Big Bug district]: Mining Jour. [Phoenix, Ariz.], v. 2, no. 11, p. 30.

————1924, Arizona News notes [San Francisco district]: Arizona Mining Jour., v. 8, no. 3, p. 30.

Bancroft, H. H., 1889, History of Arizona and New Mexico, 1530–1888: San Francisco, Calif., History Co. Pub., 829 p.
Chap. 23 considers Arizona industries including mining—early operations and gold placers.

Bancroft, Howland, 1911, A reconnaissance of the ore deposits in northern Yuma County, Arizona: U.S. Geol. Survey Bull. 451, 130 p.

Berger, H. W., 1932, Saving gold in cemented gravels [La Cholla placers]: Mining Jour. [Phoenix, Ariz.] v. 16, no. 14, p. 7, 28.

Bird, A. T., 1916, Resources of Santa Cruz County: Arizona Bur. Mines Bull. 29, 27 p.

Blacet, P. M., 1968, Geologic map of the SE ¼ Mount Union quadrangle, Yavapai County, Arizona [Turkey Creek placers]: U.S. Geol. Survey open-file map.

————1969, Lode and placer gold deposits, Mohave County, Arizona [Lost Basin and Gold Basin districts]: U.S. Geol. Survey Circ. 621, p. 12.

Black, J. A., 1890, Arizona, the land of sunshine and silver, health and prosperity; the place for ideal homes: Tucson, Ariz., Commissioner of Immigration, 143 p.
Brief descriptions of early placer-mining activity.

Blake, W. P., 1898, Remains of a species of Bos in the Quaternary of Arizona [Greaterville district]: Am. Geologist, v. 22, p. 65–72.

——1899, Historical sketch of mining in Arizona, *in* Report of the Governor of Arizona to the Secretary of the Interior: Washington, p. 43–153.

Concise historical sketch of mining in Arizona. Occurrence of gold, gold veins, mines and placers (p. 54–75) are discussed separately from silver, copper, and other mineral commodities. Many placer districts are briefly described.

Blandy, J. F., 1897, The mines of Yavapai County, Arizona: Eng. and Mining Jour., v. 63, p. 632–634.

Bray, J. C., 1933, The Bray straightline placer machine [Dos Cabezas district]: Mining Jour. [Phoenix, Ariz.], v. 17, no. 5, p. 5.

Browne, J. R., 1868, Report of J. Ross Browne on the mineral resources of the States and Territories west of the Rocky Mountains: Washington, U.S. Treasury Dept., 674 p.

Describes early placer mining history and activities in many districts throughout the State.

Bryan, Kirk, 1925, The Papago country, Arizona: U.S. Geol. Survey Water-Supply Paper 499, 436 p.

Burchard, H. C., 1882, Report of the Director of the Mint upon the statistics of the production of the precious metals in the United States (for the year 1881) : Washington, U.S. Bur. Mint, 765 p. [Arizona, p. 249–326].

——1883, Report of the Director of the Mint upon the statistics of the production of the precious metals in the United States (for the year 1882) : Washington, U.S. Bur. Mint, 873 p. [Arizona, p. 270–338].

——1884, Report of the Director of the Mint upon the production of the precious metals in the United States during the calendar year 1883: Washington, U.S. Bur. Mint, 858 p. [Arizona, p. 36–114].

——1885, Report of the Director of the Mint upon the production of the precious metals in the United States during the calendar year 1884: Washington, U.S. Bur. Mint, 644 p. [Arizona, p. 30–63].

Burgess, J. D., 1903, Recent discoveries in Arizona [Cañada del Oro district]: Eng. and Mining Jour., v. 76, p. 936.

Carter, T. L., 1911, Gold placers in Arizona [San Domingo district]: Eng. and Mining Jour., v. 91, p. 561–562.

——1912, Gold placers of Arizona—Dry washings of value [San Domingo district]: Mining and Sci. Press, v. 105, p. 166–168.

Church, J. A., 1887, Report of the Director of the Mint upon the production of the precious metals in the United States during the calendar year 1886: Washington, U.S. Treasury Dept., 343 p. [Arizona, p. 139–147].

Creasey, S. C., 1967, General geology of the Mammoth quadrangle, Pinal County, Arizona: U.S. Geol. Survey Bull. 1218, 94 p.

De Wolf, W. P., 1916, Revival of placer operation in Yavapai County, Arizona: Mining and Eng. World, v. 44, p. 199–200.

Dinsmore, C. A., 1911a, The San Domingo placers: Mining and Eng. World, v. 35, p. 793.

——1911b, Touring the mining section of the Southwest: Mining World, v. 35, p. 151–152.

Doman, R. S., 1922, The lure of Arizona gold was inspiration for Oatman: Mining Jour. [Phoenix, Ariz.], v. 6, no. 14, p. 3–4.

Drewes, Harald, 1970, Geochemical anomalies, Greaterville mining district, Arizona: U.S. Geol. Survey Bull. 1312–A, 49 p.

Elliott, W. W. (publisher), 1884, A history of the Arizona Territory: Flagstaff, Ariz., Northland Press [facsimile reprint], 322 p.
Early placer-mining history.

Ellis, E. K., 1962, Arizona gold map: Phoenix, Ariz., privately pub. by Arizona Gold Map Inc., scale 1:1,000,000.

Elsing, M. J., and Heineman, R. E. S., 1936, Arizona metal production: Arizona Bur. Mines Bull. 140, 112 p.
Tables of metal production for most mines and some placer districts.

Engineering and Mining Journal, 1890, The Walnut Grove Dam disaster in Arizona: Eng. and Mining Jour., v. 49, p. 244–245.

————1893, General mining news—Arizona [Pinto Creek]: Eng. and Mining Jour., v. 56, p. 325.

————1931, To Work Dos Cabezas placer: Eng. and Mining Jour., v. 132, p. 509.

————1933a, Trends and developments in the industry—Arizona [Copper Basin district]: Eng. and Mining Jour., v. 134, p. 87.

————1933b, Trends and developments in the industry—Arizona [Kingman area and Gold Basin district]: Eng. and Mining Jour., v. 134, p. 87.

————1933c, Trends and developments in industry—Arizona [Hassayampa River drainage area]: Eng. and Mining Jour., v. 134, p. 174.

————1933d, Trends and developments in the industry—Arizona [Kingman area]: Eng. and Mining Jour., v. 134, p. 216.

————1941, News of the industry—Arizona [Lost Basin district]: Eng. and Mining Jour., v. 142, no. 11, p. 78.

————1961, Dry land dredge extracts desert gold [San Domingo district]: Eng. and Mining Jour., v. 162, no. 8, p. 79.

Erickson, R. C., 1968, Geology and geochronology of the Dos Cabezas Mountains, Cochise County, Arizona: Southern Arizona Guidebook 3, p. 192–198.

Esenwein, William, 1958, I found gold on Rich Hill: Desert Mag., v. 21, no. 1, p. 27–29.

Farish, T. E., 1915a, History of Arizona, v. 1: Phoenix, Ariz., privately printed, 392 p.

————1915b, History of Arizona, v. 2: Phoenix, Ariz., privately printed, 348 p.
Early history of placer mining in the State.

Fickett, F. W., 1911, Dry placer mining in Quijotoa district [Arizona-Mexico]: Mining and Sci. Press, v. 102, p. 291.

Gardner, E. D., and Allsman, P. T., 1938, Power shovel and dragline placer mining: U.S. Bur. Mines Inf. Circ. 7013, 68 p.

Gardner, E. D., and Johnson, C. H., 1934, Placer mining in the Western United States, part 1: U.S. Bur. Mines Inf. Circ. 6786, 73 p.

————1935, Placer mining in the Western United States, part 3: U.S. Bur. Mines Inf. Circ. 6788, 81 p.

Gilluly, James, 1946, The Ajo mining district, Arizona: U.S. Geol. Survey Prof. Paper 209, 112 p.

————1956, General geology of central Cochise County, Arizona: U.S. Geol. Survey Prof. Paper 281, 169 p.

Girand, J. B., 1932, Centrifugal separator for recovery of placer gold [Walnut Grove and Copper Basin districts]: Mining Jour. [Phoenix, Ariz.], v. 16, no. 12, p. 3–4.

————1933, Centrifugal concentration of placer gravel [Oro Blanco district]: Eng. and Mining Jour., v. 134, p. 110–111.

Guiteras, J. R., 1936, Gold mining and milling in the Black Canyon area, Yavapai County, Arizona: U.S. Bur. Mines Inf. Circ. 6905, 51 p.

Hafer, Claud, 1911, Vulture mine and others in the Hassayampa [Maricopa County]: Mining World, v. 34, p. 1233–1234.

Hamilton, Patrick, 1884, Resources of Arizona [3d ed.]: San Francisco, Calif., Bancroft & Co. Printers, 414 p.

The history of Arizona from the discovery by Spaniards through the period of territorial settlement. Chapters on all aspects of life at that time. Chapter on mines and mining gives not only location and production but history of mining. Colorful account of life in Arizona in 1800's.

Hedburg, Edward, 1909, The Calizona placers, Arizona [Chemehuevis district]: Mining World, v. 31, p. 138.

Heikes, V. C., and Yale, C. G., 1913, Dry placers in Arizona, Nevada, New Mexico, and California: U.S. Geol. Survey Mineral Resources [1912], pt. 1, p. 254–263.

Placer-mining activity and operations in eight districts; contains information on grades of gravel and production.

Heineman, R. E. S., 1931, An Arizona gold nugget of unusual size [Weaver district]: Am. Mineralogist, v. 16, no. 6, p. 267–269.

Hill, J. M., 1910, Notes on the placer deposits of Greaterville, Arizona: U.S. Geol. Survey Bull. 430, p. 11–22.

Hinton, R J., 1878, Handbook to Arizona with appendices: San Francisco, Calif., Payot, Upham & Co., 431 p.

Description of many areas of Arizona with special references to natural resources. Placers are discussed in relation to areas of occurrence. Tables give production statistics of mines and placers of Yavapai, Pima, Mohave, Pinal, and Maricopa Counties.

Hodge, H. G., 1877, Arizona as it is; or the Coming Country: New York, Hurd & Houghton, 273 p.

History of Arizona compiled from travels during the years 1874–76.

Jagger, T. A., Jr., and Palache, Charles, 1905, Description of the Bradshaw Mountains quadrangle [Arizona]: U.S. Geol. Survey Geol. Atlas, Folio 126, 11 p.

Jahns, R. H., 1952, Pegmatite deposits of the White Picacho district, Maricopa and Yavapai Counties, Arizona: Arizona Bur. Mines Bull. 162, 105 p.

Jakosky, J. J., and Wilson, C. H., 1934, Geophysical studies in placer and water supply problems [Pinal County]: Am. Inst. Mining Metall. Engineers Tech. Pub. 515, p. 3–18.

Johnston, W. P., and Lowell, J. D., 1961, Geology and origin of mineralized breccia pipes in Copper Basin, Arizona: Econ. Geology. v. 56, p. 916–940.

Jones, E. L., Jr., 1915, Gold deposits near Quartzsite, Arizona: U.S. Geol. Survey Bull. 620, p. 45–57.

———1916, A reconnaissance in the Kofa Mountains, Arizona: U.S. Geol. Survey Bull. 620, p. 151–164.

Keiser, W. G., 1916, Dry placer mining on a large scale [Plomosa district]: Mining and Eng. World, v. 44, p. 999–1000.

Konselman, A. S., 1933, Federal grubstaking of placer mines: Mining Jour. [Phoenix, Ariz.], v. 17, no. 12, p. 5.

Koschmann, A. H., and Bergendahl, M. H., 1968, Principal gold-producing districts in the United States: U.S. Geol. Survey Prof. Paper 610, 283 p.

General summary of 42 important gold districts in Arizona. Main emphasis is on lode mines; information about placers is taken mostly from Wilson (1952, 5th ed.) .

Krieger, M. H., 1965, Geology of the Prescott and Paulden quadrangles, Arizona: U.S. Geol. Survey Prof. Paper 427, 127 p.

Land, George, 1931, Gold at Dos Cabezas: Mining Jour. [Phoenix, Ariz.], v. 14, no. 24, p. 28.

Lausen, Carl, 1931, Geology and ore deposits of the Oatman and Katherine districts, Arizona: Arizona Bur. Mines Bull. 131, 126 p.

Lausen, Carl, and Wilson, E. D., 1925, Gold and copper deposits near Payson, Arizona: Arizona Bur. Mines Bull. 120, 44 p.

Lindgren, Waldemar, 1905a, The copper deposits of the Clifton-Morenci district, Arizona: U.S. Geol. Survey Prof. Paper 43, 375 p.

———1905b, Description of the Clifton quadrangle [Arizona]: U.S. Geol. Survey Geol. Atlas, Folio 129, 14 p.

———1926, Ore deposits of the Jerome and Bradshaw Mountain quadrangles, Arizona: U.S. Geol. Survey Bull. 782, 192 p.

MacHunter, Audrey, and Henderson, Randall, 1958, Boom days in old La Paz: Desert Mag., v. 21, no. 9, p. 19–21.

Maltman, A., 1917, Dry placer mining [Plomosa district]: Mining and Sci. Press, v. 114, p. 203.

Maynard, G. W., 1907, Examining a placer property [Greaterville district]: Mining World, v. 27, p. 7–8.

McConnell, A. B., 1911, Arizona: Mining World, v. 34, p. 143–147.

Mining and Scientific Press, 1907, General mining news—Arizona [Barbarossa-Dripping Spring district]: Mining and Sci. Press, v. 94, p. 677.

———1908, General mining news—Arizona [Patagonia district]: Mining and Sci. Press, v. 96, p. 649.

———1916, The mining summary, Yuma County: Mining and Sci. Press, v. 113, p. 249.

Mining Journal, 1932a, Concentrates from the Southwest—Arizona [Copper Basin district]: Mining Jour. [Phoenix, Ariz.], v. 15, no. 23, p. 18.

———1932b, Concentrates from the Southwest—Arizona [San Francisco district]: Mining Jour. [Phoenix, Ariz.], v. 15, no. 24, p. 18.

———1932c, Nuggets from the Western States—Arizona [Chemehuevis district]: Mining Jour. [Phoenix, Ariz.], v .16, no. 14, p. 16.

———1933, Nuggets from the Western States—Arizona [Lost Basin district]: Mining Jour. [Phoenix, Ariz.], v. 17, no. 5, p. 10.

———1938a, Concentrates from Western States—Arizona [Baboquivari district]: Mining Jour. [Phoenix, Ariz.], v. 22, no. 4, p. 18.

———1938b, Concentrates from Western States—Arizona [San Francisco River placers]: Mining Jour. [Phoenix, Ariz.], v. 22, no. 14, p. 19.

———1938c, Dry separation process in use by Universal Placer Corporation: Mining Jour. [Phoenix, Ariz.], v. 21, no. 24, p. 42–43.

———1938d, Mill heads from the Western States—Arizona [Gila River placers]: Mining Jour. [Phoenix, Ariz.], v. 22, no. 6, p. 19.

———1938e, Mill heads from the Western States—Arizona [Weaver district]: Mining Jour. [Phoenix, Ariz.], v. 21, no. 24, p. 20.

———1939a, Concentrates from Western States—Arizona [La Cholla placers]: Mining Jour. [Phoenix, Ariz.] v. 22, no. 23, p. 19.

———1939b, Mill heads from the Western States—Arizona [Payson district]: Mining Jour. [Phoenix, Ariz.], v. 23, no. 10, p. 17.

———1939c, Mill heads from the Western States—Arizona [Quijotoa district]: Mining Jour. [Phoenix, Ariz.], v. 23, no. 13, p. 17.

———1940, Mill heads from the Western States—Arizona [Quijotoa district]: Mining Jour. [Phoenix, Ariz.], v. 24, no. 1, p. 17.

———1941, Concentrates from Western States—Arizona [San Pablo placer]: Mining Jour. [Phoenix, Ariz.], v. 24, no. 2, p. 20.

——1946, Concentrates from Western States [Comobabi district]: Mining Jour. [Phoenix, Ariz.], v. 29, no. 19, p. 16.

Mining Reporter, 1906, Revival of gold mining in the Clifton-Morenci, Arizona, district: Mining Reporter, v. 53, p. 386–388.

Mining Review, 1910a, Colorado River placers: Mining Review [Salt Lake City, Utah], v. 12, no. 10, p. 33.

——1910b, Nogales gold excitement: Mining Review [Salt Lake City, Utah], v. 12, no. 16, p. 33.

Mining World, 1909, Late news from the worlds mining camps—Arizona [Ellsworth district]: Mining World, v. 30, p. 993.

——1911, Late news from the worlds mining camps—Arizona [Huachuca district]: Mining World, v. 34, p. 797.

Moolick, R. T., and Durek, J. J., 1966, The Morenci district, in Titley, S. R., and Hicks, C. L., eds., Geology of the porphyry copper deposits, southwestern North America: Tucson, Ariz., Univ. Arizona Press, p. 221–231.

Moore, R. T., 1969, Gold, in Mineral and water resources of Arizona: U.S. 90th Cong., 2d sess., Comm. Interior and Insular Affairs, Comm. Printing, p. 156–167.
Brief summary of history of placer mining.

Mowry, Sylvester, 1863, The geography and resources of Arizona and Sonora: San Francisco, Calif., and New York, A. Roman & Co., 124 p.
Early placer-mining history.

Nolan, T. B., 1936, Nonferrous metal deposits, in Hewett, D. F., and others, Mineral resources of the region around Boulder Dam: U.S. Geol. Survey Bull. 871, p. 5–77.
Summarizes gold, silver, and copper deposits for counties of Arizona on p. 10–34. Briefly locates some important placer deposits.

Peterson, N. P., 1962, Geology and ore deposits of the Globe-Miami district, Arizona: U.S. Geol. Survey Prof. Paper 342, 151 p.

Peterson, N. P., Gilbert, C. M., and Quick, G. L., 1951, Geology and ore deposits of the Castle Dome area, Gila County, Arizona: U.S. Geol. Survey Bull. 971, 134 p.

Plummer, W. L. 1916, Successful dry placer operations at Plumosa, Arizona: Mining and Eng. World, v. 45, p. 1–3.

Randolph, C. C., 1901, Report to the Director of the Mint upon the production of the precious metals in the United States during the calendar year 1900: Washington, U.S. Bur. Mint, 380 p. [Arizona, p. 71–80].

——1903, Report of the Director of the Mint upon the production of the precious metals in the United States during the calendar year 1902: Washington, U.S. Bur. Mint, 391 p. [Arizona, p. 55–75].

Ransome, F. L., 1903, Geology of Globe copper district, Arizona: U.S. Geol. Survey Prof. Paper 12, 168 p.

——1904a, Description of the Bisbee quadrangle [Arizona]: U.S. Geol. Survey Geol. Atlas, Folio 112, 17 p.

——1904b, Description of the Globe quadrangle [Arizona]: U.S. Geol. Survey Geol. Atlas, Folio 111, 17 p.

——1904c, Geology and ore deposits of the Bisbee quadrangle, Arizona: U.S. Geol. Survey Prof. Paper 21, 168 p.

——1923a, Description of the Ray quadrangle [Arizona]: U.S. Geol. Survey Geol. Atlas, Folio 217, 23 p.

——1923b, Geology of the Oatman gold district, Arizona, a preliminary report: U.S. Geol. Survey Bull. 743, 58 p.

Raymond, R. W., 1872, Statistics of mines and mining in the States and Territories West of the Rocky Mountains for the year 1870: Washington, D.C., U.S. Treasury Dept., 566 p. [Arizona, p. 224–281].

———1874, Statistics of mines and mining in the States and Territories west of the Rocky Mountains, being the sixth annual report (for the year 1873) ; Washington, U.S. Treasury Dept., 585 p. [Arizona, p. 344–350].

———1875, Statistics of mines and mining in the States and Territories west of the Rocky Mountains, being the seventh annual report for the year 1874: Washington, U.S. Treasury Dept., 540 p. [Arizona, p. 389–395].

———1877, Statistics of mines and mining in the States and Territories west of the Rocky Mountains, being the eighth annual report (for the year 1875) : Washington, U.S. Treasury Dept., 519 p. [Arizona, p. 341–354].

Robinson, R. F., and Cook, Annan, 1966, The Safford copper deposit, Lone Star mining district, Graham County, Arizona, in Titley, S. R., and Hicks, C. L., eds., Geology of the porphyry copper deposits, southwestern North America: Tucson, Ariz., Univ. Arizona Press, p. 221–231.

Root, W. A., 1912, The dry placers of Yuma County, Arizona: Mining and Eng. World, v. 36, p. 758.

———1915, Dredge mining operations in Santa Rita Mountains, Arizona: Mining and Eng. World, v. 42, p. 377.

Roseveare, G. H., 1961, Appendix—operations during 1951–1961, in Wilson, E. D., Gold placer and placering in Arizona: Arizona Bur. Mines Bull. 168, p. 119.

Salt Lake Mining Review, 1923, In nearby states—Arizona [San Francisco district]: Salt Lake Mining Rev. v. 25, no. 18, p. 23.

Sawyer, D. L., 1932, Sampling a gold placer [Weaver district]: Eng. and Mining Jour., v. 133, p. 381–383.

Schrader, F. C., 1909, Mineral deposits of the Cerbat Range, Black Mountains and Grand Wash Cliffs, Mohave County, Arizona: U.S. Geol. Survey Bull. 297, 226 p.

———1915, Mineral deposits of the Santa Rita and Patagonia Mountains, Arizona: U.S. Geol. Survey Bull. 582, 373 p.

Schrader, F. C., Stone, R. W., and Sanford, S., 1917, Useful minerals of the United States: U.S. Geol. Survey Bull. 624, 412 p.
Gold placer districts in Arizona are listed on page 24.

Sparks, G. M., 1917, Yavapai, the land of opportunity: Arizona Mining Jour., v. 1, no. 4, p. 1–4.

Stephens, B. A., 1884, Quijotoa mining district guidebook: Tucson, Ariz., Tucson Citizen, Printing & Pub. Co.

Stipp, T. F., Jaigler, L. B., Alto, B. R., and Sutherland, H. L., 1967, Reported occurrences of selected minerals in Arizona: U.S. Geol. Survey Mineral Inv. Resource Map MR–46, scale 1:500,000.

Tovote, W. L., 1910, The Clifton-Morenci district of Arizona, part 1: Mining and Sci. Press, v. 101, p. 770–773.

Trippel, Alex, 1888, Report of the Director of the Mint upon production of the precious metals in the United States during the calendar year 1887: Washington, U.S. Bur. Mint, 375 p. [Arizona, p. 120–142].

———1889, Report of the Director of the Mint upon production of the precious metals in the United States during the calendar year 1888: Washington, U.S. Bur. Mint, 246 p. [Arizona, p. 88–91].

Trischka, Carl, 1938, Bisbee district, in Some Arizona ore deposits: Arizona Bur. Mines Bull. 145, p. 32–41.

U.S. Bureau of Mines, 1925–34, Mineral resources of the United States [annual volumes, 1924–31]: Washington, U.S. Govt. Printing Office.

——1933–68, Minerals Yearbook [annual volumes, 1932–68]: Washington, U.S. Govt. Printing Office.
Information relating to placers cited in text is referenced by year of pertinent volume.

——1967, Production potential of known gold deposits in the United States: U.S. Bur. Mines Inf. Circ. 8331, 24 p.
Gives estimates of total placer-gold production in troy ounces.

U.S. Geological Survey, 1896–1900, Annual reports [17th through 21st, 1895–1900]: Washington, U.S. Govt. Printing Office.

——1883–1924, Mineral resources of the United States [annual volumes, 1882–1923]: Washington, U.S. Govt. Printing Office.
Information relating to placers cited in text is referenced by year of pertinent volume.

——1968, U.S. Geological Survey Heavy Metals Program progress report 1966 and 1967 [Lost Basin and Gold Basin districts]: U.S. Geol. Survey Circ. 560, 24 p.

Watson, H. B., 1918, Rich Hill observations: Arizona Mining Jour., v. 2, no. 7, p. 8–10, 26.

Webber, B. N., 1935, Bajada placers of the arid southwest: Am. Inst. Mining Metall. Engineers Trans., v. 115, Mining Geology, p. 378–391.
Ascribes genesis of slope deposits "Bajadas" primarily to arid conditions, wind work, and sheet floods. Compares "Bajada" placers to stream placers. E. D. Wilson's discussion (p. 391) emphasizes work of water in rills and channels and derivation of placers from small gold-bearing veins.

Weber, R. H., 1948, Geology of the east-central portion of the Huachuca Mountains, Arizona [abs.]: Geol. Soc. America Bull. 59, no. 12, p. 1384–1385.

Willis, C. F., 1915, Las Guijas placer: Arizona, the State Mag. and Pathfinder, Nov. 1915, p. 10.

——1916a, Arizona: Arizona Bur. Mines Bull. 6, 16 p.

——1916b, Mining in Arizona: Mining and Sci. Press, v. 112, p. 299–300.
General state of mining in Arizona at that time including a location map of 15 well-known placer districts.

Wilson, E. D., 1927, Geology and ore deposits of the Courtland-Gleeson region, Arizona: Arizona Bur. Mines Bull. 123, 79 p.

——1933, Geology and mineral deposit of southern Yuma County, Arizona: Arizona Bur. Mines Bull. 134, 234 p.

——1941, Tungsten deposits of Arizona: Arizona Bur. Mines Bull. 148, 54 p.

——1951, Arizona lead and zinc deposits: Arizona Bur. Mines Bull. 158, 115 p.

——1961, Gold placers and placering in Arizona [6th ed.], revised: Arizona Bur. Mines Bull. 168, 124 p.; preceded by: 1927, 2d ed., revised, Bull. 124, 60 p.; 1932, Bull. 132, pt. 1, p. 1–71; 1933, 4th ed., Bull. 135, 90 p.; 1937, 4th ed., revised, Bull. 142, 90 p.; 1952, 5th ed., revised, Bull. 160, pt. 1, p. 11–68.
Series of bulletins that describe the location, extent, history, placer-mining activity, and production of most Arizona placers. Early bulletins are virtually identical with the sixth edition. Bulletins describe details of mining activity during the 1932–33 mining season, when most of the districts described were visited. Production data in text have been revised for each reissue of the bulletin up to the sixth edition.

——1962, A résumé of the geology of Arizona: Arizona Bur. Mines Bull. 171, 140 p.
Summarizes age of bedrock and of lode deposits; history of mining.

Wilson, E. D., Cunningham, J. B., and Butler, G. M., 1934, Arizona lode gold mines and gold mining: Arizona Bur. Mines Bull. 137, 261 p.

Wilson, E. D., O'Haire, R. T., and McCrory, F. J., 1961, Map and index of Arizona mining districts: Arizona Bur. Mines, scale 1:1,000,000.

Wood, H. R., 1929, History of mining in Yavapai County, Arizona: Mining Jour. [Phoenix, Ariz.], v. 13, no. 8, p. 9, 35–36.

GEOLOGIC MAP REFERENCES

[References keyed by number to districts given in text]

Anderson, C. A., and Blacet, P. M., 1972a, Geologic map of the Mayer quadrangle, Yavapai County, Arizona: U.S. Geol. Survey Geol. Quad. Map GQ–996, scale 1:62,500.
Nos. 57–59.

———1972b, Geologic map of the Mount Union quadrangle, Yavapai County, Arizona: U.S. Geol. Survey Geol. Quad. Map GQ–997, scale 1:62,500.
Nos. 55–58.

Anderson, C. A., Scholz, E. A., and Strobell, J. D., Jr., 1955, Geology and ore deposits of the Bagdad area, Yavapai County, Arizona: U.S. Geol. Survey Prof. Paper 278, pl. 3.
No. 65.

Arizona Bureau of Mines, 1958, Geologic map of Yavapai County, Arizona: Tucson, Ariz., Bur. Mines, scale 1:375,000.
Nos. 56, 59–63, 65.

Cooper, J. R., 1960, Reconnaissance map of the Wilcox, Fisher Hills, Cochise, and Dos Cabezas quadrangles, Cochise and Graham Counties, Arizona: U.S. Geol. Survey Mineral Inv. Field Studies Map MF–231, scale 1:62,500.
No. 1.

Creasey, S. C., 1967, General geology of the Mammoth quadrangle, Pinal County, Arizona: U.S. Geol. Survey Bull. 1218, 94 p., pl. 1.
No. 46.

Drewes, Harald, 1971a, Geologic map of the Mount Wrightson quadrangle, Santa Cruz and Pima Counties, Arizona: U.S. Geol. Survey Misc. Geol. Inv. Map I–614, scale 1:48,000.
Nos. 36, 54.

———1971b, Geologic map of the Sahuarita quadrangle, Pima County, Arizona: U.S. Geol. Survey Misc. Geol. Inv. Map I–613, scale 1:48,000.
No. 36.

Gilluly, James, 1946, The Ajo mining district, Arizona: U.S. Geol. Survey Prof. Paper 209, 112 p., pls. 3, 20, 21.
No. 42.

———1956, General geology of central Cochise County, Arizona: U.S. Geol. Survey Prof. Paper 281, 169, p., pl. 5.
No. 2.

Hayes, P. T., and Landis, E. R., 1964, Geologic map of the southern part of the Mule Mountains, Cochise County, Arizona: U.S. Geol. Survey Misc. Geol. Inv. Map I–418, scale 1:48,000.
No. 3.

Hayes, P. T., and Raup, R. B., 1968, Geologic map of the Huachuca and Mustang Mountains, southeastern Arizona: U.S. Geol. Survey Misc. Geol. Inv. Map I–509, scale 1:48,000.
No. 4.

Heindl, L. A., and McCullough, R. A., 1961, Geology and the availability of water in the lower Bonita Creek area, Graham County, Arizona: U.S. Geol. Survey Water-Supply Paper 1589, 56 p., pl. 1.
No. 15.

Hill, J. M., 1910, Notes on the placer deposits of Greaterville, Arizona: U.S. Geol. Survey Bull. 430, p. 11–22 and sketch map of placer deposits.
No. 36.

Jones, E. L., Jr., 1915, Gold deposits near Quartzsite, Arizona: U.S. Geol. Survey Bull. 620, pl. 4.
No. 77.

——1916, A reconnaissance in the Kofa Mountains, Arizona: U.S. Geol. Survey Bull. 620, pl. 5.
No. 74.

Krieger, M. H., 1965, Geology of the Prescott and Paulden quadrangles, Arizona: U.S. Geol. Survey Prof. Paper 427, 127, p., pls. 1, 2.
Nos. 55, 64

——1967, Reconnaissance geologic map of the Iron Springs quadrangle, Yavapai County, Arizona: U.S. Geol. Survey Misc. Geol. Inv. Map I–504.
No. 63.

Lausen, Carl, 1931, Geology and ore deposits of the Oatman and Katherine districts, Arizona: Arizona Bur. Mines Bull. 131, pl. 1.
No. 28.

Lausen, Carl, and Wilson, E. D., 1925, Gold and copper deposits near Payson, Arizona: Arizona Bur. Mines Bull. 120, pl. 1.
No. 8.

Lindgren, Waldemar, 1905a, The copper deposits of the Clifton-Morenci district, Arizona: U.S. Geol. Survey Prof. Paper 43, pl. 1.
Nos. 13, 14.

——1926, Ore deposits of the Jerome and Bradshaw Mountains quadrangles, Arizona: US. Geol. Survey Bull. 782, pl. 2.
No. 60.

Longwell, C. R., 1936, Geology of the Boulder Reservoir floor, Arizona-Nevada: Geol. Soc. America Bull., v. 47, p. 1393–1476, pls. 2–4.
No. 30.

——1963, Reconnaissance geology between Lake Mead and Davis Dam, Arizona-Nevada: U.S. Geol. Survey Prof. Paper 374–E, pl. 1.
No. 30.

Miller, F. K., 1970, Geologic map of the Quartzsite quadrangle, Yuma County, Arizona: U.S. Geol. Survey Geol. Quad. Map GQ–841, scale 1:62,500.
No. 79.

Peterson, D. W., 1960, Geologic map of the Haunted Canyon quadrangle: U.S. Geol. Survey Geol. Quad. Map GQ–128, scale 1:24,000.
No. 9.

Peterson, N, P., 1962, Geology and ore deposits of the Globe-Miami district, Arizona: U.S. Geol. Survey Prof. Paper 342, 151 p., pl. 1. (Lost Gulch, Gold Gulch, Pinto Creek, and Pinal Creek).
No. 9.

Peterson, N. P., Gilbert, C. M., and Quick, G. L., 1951, Geology and ore deposits of the Castle Dome area, Gila County, Arizona: U.S. Geol. Survey Bull. 971, 134 p., pl. 1. (Gold Gulch.)
No. 9.

Ransome, F. L., 1904, Description of the Globe quadrangle [Arizona]: U.S. Geol. Survey Geol. Atlas, Folio 111, 17 p.

No. 9.

————1923a, Description of the Ray quadrangle [Arizona]: U.S. Geol. Survey Geol. Atlas, Folio 217, 23 p.

No. 10.

————1923b, Geology of the Oatman gold district, Arizona: U.S. Geol. Survey Bull. 743, pl. 1.

No. 28.

Schrader, F. C., 1909, Mineral deposits of the Cerbat Range, Black Mountains and Grand Wash Cliffs, Mohave County, Arizona: U.S. Geol. Survey Bull. 297, pl. 1.

Willden, Ronald, 1964, Geology of the Christmas quadrangle, Gila and Pinal Counties, Arizona: U.S. Geol. Survey Bull. 1161-E, pl. 1.

No. 10.

Wilson, E. D., 1960, Geologic map of Yuma County, Arizona: Arizona Bur. Mines, scale 1:375,000.

Nos. 69–78.

Wilson, E. D., and Moore, R. T., 1958, Geologic map of Graham and Greenlee Counties, Arizona: Arizona Bur. Mines, scale 1:375,000.

Nos. 13–15.

————1959a, Geologic map of Mohave County: Arizona Bur. Mines, scale 1:375,000.

Nos. 27–31.

————1959b, Geologic map of Pinal County, Arizona: Arizona Bur. Mines, scale 1:375,000.

No. 46.

Wilson, E. D., Moore, R. T., and Cooper, J. R., 1969, Geologic map of Arizona: U.S. Geol. Survey, scale 1:500,000.

Wilson, E. D., Moore, R. T., and O'Haire, R. T., 1960, Geologic map of Pima and Santa Cruz Counties, Arizona: Arizona Bur. Mines, scale 1:375,000.

Nos. 35, 37–41, 51–54.

Wilson, E. D., Moore, R. T., and Peirce, H. W., 1957, Geologic map of Maricopa County, Arizona: Arizona Bur. Mines, scale 1:375,000.

Nos. 17–20.

————1959, Geologic map of Gila County, Arizona: Arizona Bur. Mines, scale 1:375,000.

No. 8.

www.ingramcontent.com/pod-product-compliance
Lightning Source LLC
Chambersburg PA
CBHW032009190326
41520CB00007B/412